Simulations for Digital Electronics Using Electronics Workbench®

James L. Antonakos
Broome Community College

Prentice Hall
Upper Saddle River, New Jersey Columbus, Ohio

Editor: Scott Sambucci
Production Editor: Rex Davidson
Design Coordinator: Karrie M. Converse
Cover Designer: Brian Deep
Production Manager: Patricia A. Tonneman
Marketing Manager: Ben Leonard

This book was printed and bound by Banta Company/Harrisonburg. The cover was printed by Phoenix Color Corp.

Printed in the United States of America

10 9 8 7 6 5 4 3 2 1

ISBN: 0-13-646423-8

Prentice-Hall International (UK) Limited, *London*
Prentice-Hall of Australia Pty. Limited, *Sydney*
Prentice-Hall Canada Inc., *Toronto*
Prentice-Hall Hispanoamericana, S. A., *Mexico*
Prentice-Hall of India Private Limited, *New Delhi*
Prentice-Hall of Japan, Inc., *Tokyo*
Pearson Education Asia Pte. Ltd., *Singapore*
Editora Prentice-Hall do Brasil, Ltda., *Rio de Janeiro*

Preface

The tremendous increase in computing power over the past several years has paved the way for computer-based simulation of electronic circuitry (both analog and digital), including the use of 'virtual' instruments such as the dual-trace oscilloscope, Bode plotter, function generator, and power supply. Electronics Workbench provides these capabilities, and many more, with an easy-to-learn-and-use environment. This laboratory manual uses Electronics Workbench to simulate the operation of many typical digital electronics circuits, from combinational logic through flip flops, counters, arithmetic, synchronous logic, D/A converters, and others. Each laboratory contains an Objectives section, an Introduction section summarizing the relevant theory, a multi-step Procedure containing schematic figures for the circuits under examination, plus tables for measured data, and a Discussion section that poses additional questions for consideration.

Nothing can replace the 'hands-on' experience of actually setting up a real circuit on a lab bench. But learning how to use the virtual instruments of Electronics Workbench, including its Logic Analyzer and Word Generator, requires the same correct thinking and handiwork as in using the real instruments. Typically, instruments and test components, such as signal sources or switches, are not shown in the procedure schematics. The student is encouraged right from the beginning to learn how to correctly connect the necessary components to complete the test circuit. An instructor's guide is available that shows the proper placement and connections. Furthermore, every lab procedure contains a troubleshooting exercise. A circuit with a fault introduced into it is provided for the student to analyze. The exact nature of the fault is hidden (see the instructor's guide for solution) and could be anything you might typically find in an actual circuit, such as a shorted or open component or faulty logic gate.

A diskette is provided with this manual that contains all of the circuit files necessary to run each experiment. The student is directed to load a specific file while performing the procedure. Once the circuit file is loaded, additional components must be added to make it fully operational (such as switches to supply zero/one levels or an oscilloscope to examine the input and output waveforms). Troubleshooting circuits are password protected, to protect the identity of the fault.

The instructor may pick and choose labs as desired, and even double up labs if necessary, as some of the experiments are shorter than others. Ideally, students should work alone on each experiment, to ensure that each knows the proper ways to connect instruments and make measurements. This, of course, is left to the discretion of the instructor.

Using a computer to run lab experiments has many advantages. It costs less to purchase one computer than a full bench of equipment. The virtual instruments will not go out of calibration or blow up if connected improperly. A complicated circuit setup can be saved on disk. These are just a few of the advantages of computer simulation. Electronics Workbench makes it all possible.

James L. Antonakos
antonakos_j@sunybroome.edu
http://www.sunybroome.edu/~antonakos_j

This is for my little girl,
Ashley Michele Antonakos.

"Hey Peanut!"

Contents

Experiment 1

Introduction to Electronics Workbench

Name ______________________ **Date** __________

Objectives

The objectives of this experiment are to:

- learn how to open, modify, and save a circuit file.
- demonstrate interactive simulation capabilities.
- understand how to find things in Electronics Workbench.

Introduction

This experiment is designed to introduce Electronics Workbench to a novice. Experienced users may wish to skip ahead to Experiment 2.

Electronics Workbench is a complete electronics lab inside the computer. Using only simple mouse movements (or keyboard commands) you can set up an analog or a digital circuit (or a mixture of both), simulate its operation, make measurements with virtual instruments, and *never worry* about hooking up something wrong and causing damage.

The procedure is designed to have you work your way through a simple circuit. It is assumed that you know the following:

- how to navigate around directories.
- where your EWB files for this lab manual are stored.
- how to use the mouse inside Windows.

Procedure

1. Load the circuit **E1-1**, shown in Figure 1.1. This is accomplished by left-clicking on the 'File' menu and then left-clicking on the 'Open' option. You will get a list of EWB circuit files. If you are not in the proper directory, navigate to it and then double-click on the circuit file **E1-1** to open it.

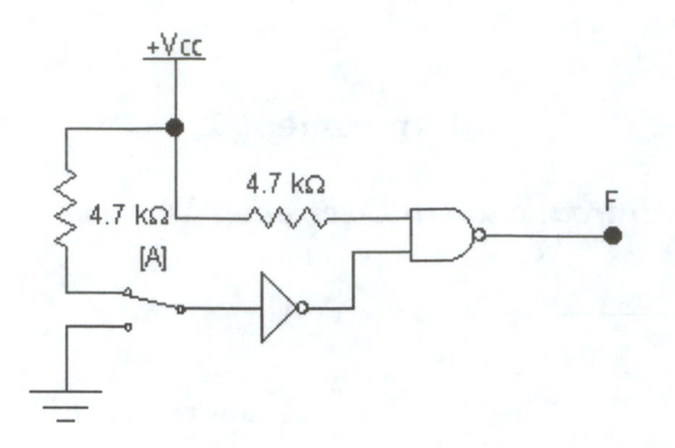

Figure 1.1: Sample circuit

2. To determine if the circuit is operating correctly, let us examine the state of the NAND gate output. You can use the logic indicator from the 'Indicators' parts bin (see Appendix A for assistance). Do this by clicking on the icon for the 'Indicators' bin. A window with various parts should open next to the workspace.

   To place a logic indicator, grab the round, red indicator (left-click and hold to grab), pull it over into the workspace, and release the left mouse button. Your circuit should now look like Figure 1.2.

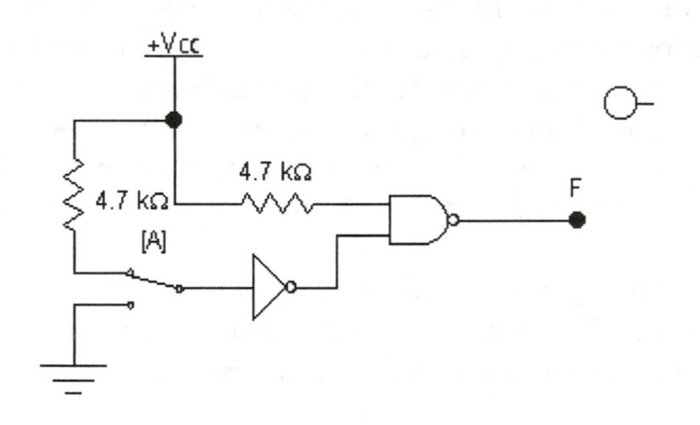

Figure 1.2: Adding a logic indicator

3. To connect the logic indicator to the F output, move the mouse pointer until it is near the round connection node labeled F. When you get close enough a small black dot will appear, indicating that you are starting to place a wire. Left-click and hold on the black dot. As you move the mouse, a thin black 'wire' should appear between the black dot and the mouse pointer. Move the mouse pointer close to the wire coming out of the logic indicator. When you get close, another black dot will appear. Release the mouse button to make the connection. You should now see something similar to Figure 1.3.

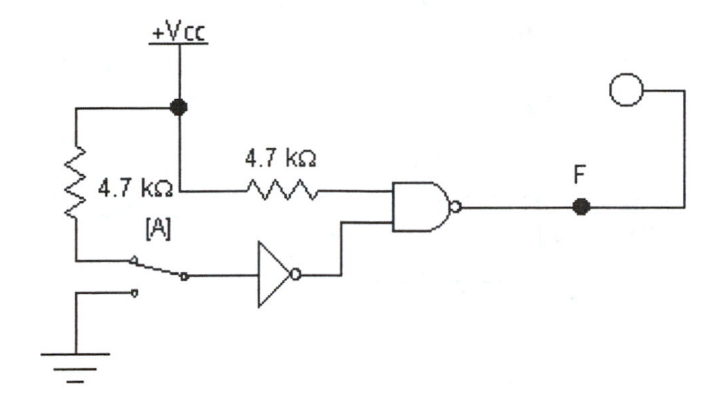

Figure 1.3: Connecting the logic indicator

The logic indicator is used instead of a LED, which requires additional time to be simulated properly. Before you connect the logic indicator, you can rotate it around to get the connection lead where you want it. Use Control-R to rotate any selected component. A component turns red when selected by a single left-click.

4. Turn the circuit on by left-clicking on the power switch. The logic indicator should glow red, indicating a logic one level. Congratulations! You have correctly installed a logic indicator.

5. The switch labeled A is controlled by pressing the A key. Each keypress moves the switch to its opposite position. The 4.7 K ohm resistor is used to create a logic one level on the output of the switch. The ground connection is for making a logic zero. Press the A key a number of times. Does the logic indicator turn on and off?

6. Now modify the circuit by adding an inverter on the output of the NAND gate. It is necessary to disconnect one end of the wire from the NAND gate output to the F node. Move the mouse pointer close to one end of the wire. Once the small black dot appears, left-click and hold, then drag the pointer away from the component. The wire should disconnect and follow your mouse pointer around. Simply releasing the left mouse button erases the wire because only one end is still connected. You should see something similar to Figure 1.4 now.

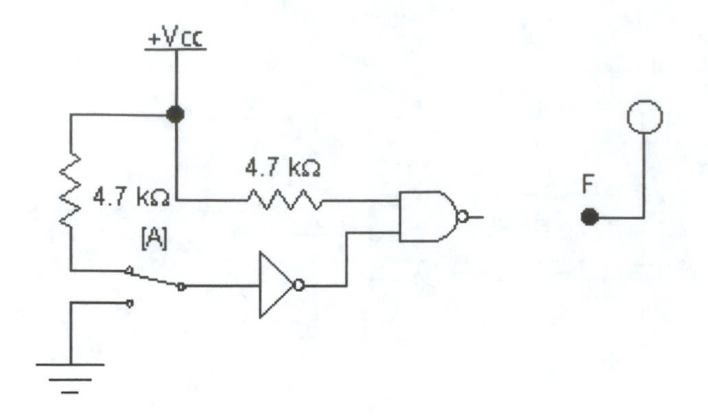

Figure 1.4: The output wire has been removed

7. Grab a second inverter from the Logic Gates bin, place it between the NAND gate and the output node, and connect it. Figure 1.5 shows the final circuit.

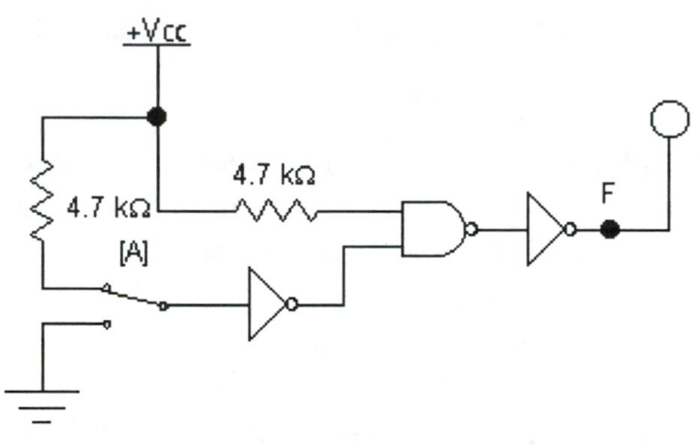

Figure 1.5: Final test circuit

8. Test the new circuit to verify its operation.

9. Once a component has been placed and connected, it is often necessary to change its value. To change the value of either 4.7 K ohm resistor, double-click on it, or single-click to select it and then go to the 'Circuit' menu and select the 'Value' option. A resistor dialog box appears that allows you to enter the new resistor value. Enter 2.7 and be sure 'K ohms' is also indicated. Then click on 'OK.'

   The same method can be used to change the keyboard key that controls the switch.

10. To save the new circuit, go to the 'File' menu and select the 'Save' or 'Save as…' option. 'Save' is usually used to simply update the current circuit file (a warning is issued that allows you to cancel the Save operation if necessary). "Save as…' is used to save the file under a different name.

11. *Troubleshooting*: Electronics Workbench allows faults to be introduced into the circuit that change the operating behavior. To practice troubleshooting skills, this lab manual contains a troubleshooting step in each experiment. Faults have been added to special troubleshooting circuits and then *hidden* so that they are not obvious. Faults are typically due to opens, shorts, or leaky components. The nature of the fault is revealed in the instructor's guide available to your instructor.

    You must develop your own skills as a troubleshooter, but these tips should come in handy:

    - Is everything connected to power and ground?
    - Are you using the instrument the right way?
    - Are all component values correct?
    - Are there any shorts or opens?
    - Are there any leaky components?

Try your skills as a troubleshooter with the **E1-2** circuit. Load it in and start looking for the fault. Figure 1.6 shows the circuit.

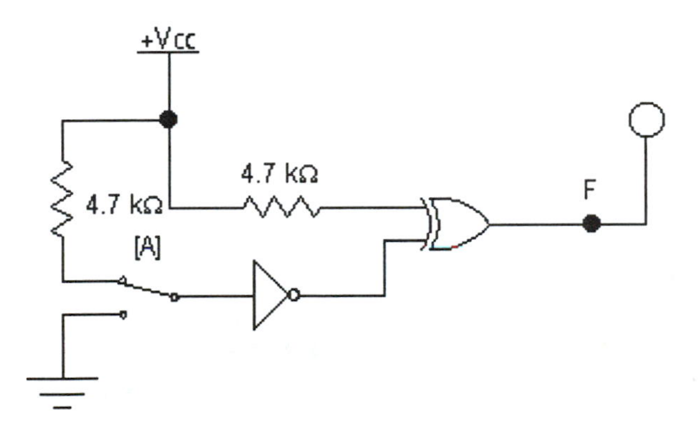

Figure 1.6: Troubleshooting circuit

Note that this is similar to the original circuit, except for a hidden fault and new logic gate. Something *must* be wrong because the logic indicator is always on during simulation. Good luck!

Discussion

While reviewing your data and results, provide detailed answers to each of the following:

1. How is a parts bin selected?

2. How is a component selected?

3. How are wires added to the circuit?

4. When is the circuit simulated?

5. Will the computer burst into flames if your circuit is connected wrong?

Experiment 2

The Word Generator and Logic Analyzer

Name ______________________ **Date** __________

Objectives

The objectives of this experiment are to:

- Examine the operation of the Word Generator and the Logic Analyzer.
- Use the Word Generator and Logic Analyzer to test a simple circuit.

Introduction

Two virtual instruments useful during digital analysis are the Word Generator and the Logic Analyzer. The Word Generator is used to apply predetermined patterns to the inputs of a digital circuit. The Logic Analyzer is used to capture and display a set of digital waveforms from various outputs of the circuit.

Figure 2.1(a) shows the Word Generator in minimized form. Figure 2.1(b) shows the expanded details of the Word Generator.

0000 XXXX

Figure 2.1(a): Word Generator

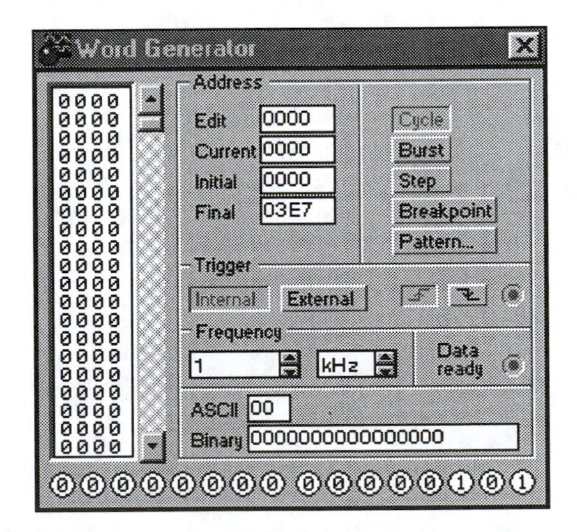

Figure 2.1(b): Word Generator details

Across the bottom are 16 outputs whose pattern corresponds to the word being output from the generator. The patterns may be entered into the Word Generator in binary, hexadecimal, or even ASCII.

Each pattern occupies a specific address in the Word Generator. Addresses are generated sequentially while the Word Generator is running and may be adjusted to limit the range of locations used during simulation.

The Cycle, Burst, and Step modes determine how data is output from the Word Generator. Cycle mode continuously outputs the selected range, Burst outputs the range of data once, and Step outputs one new pattern at a time. The Word Generator can be triggered externally by a rising or falling edge, or internally through its own clock (you can adjust the frequency). A Data Ready output is available to indicate to external circuitry when data is available.

Clicking the Pattern button brings up the window shown in Figure 2.2.

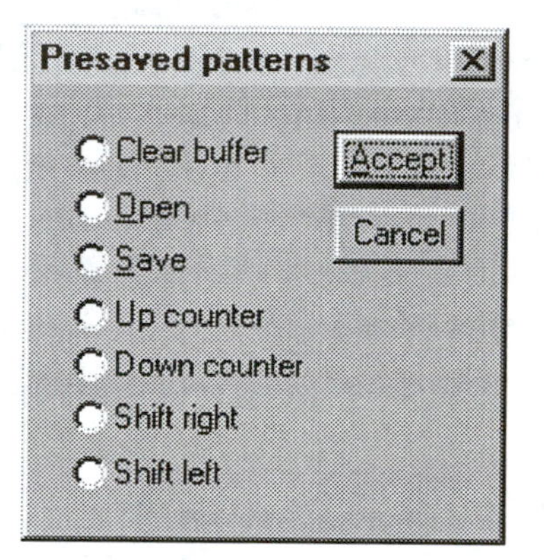

Figure 2.2: Patterns menu

Note that you may load or save patterns to disk, as well as preload the Word Generator with one of four built-in patterns.

Figure 2.3(a) shows the minimized Logic Analyzer. The left side contains 16 inputs, all of which are sampled and displayed in the expanded version of the Logic Analyzer shown in Figure 2.3(b).

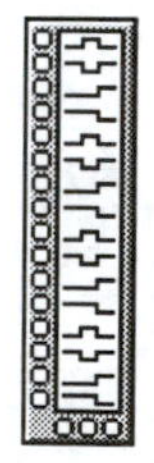

Figure 2.3(a): Logic Analyzer

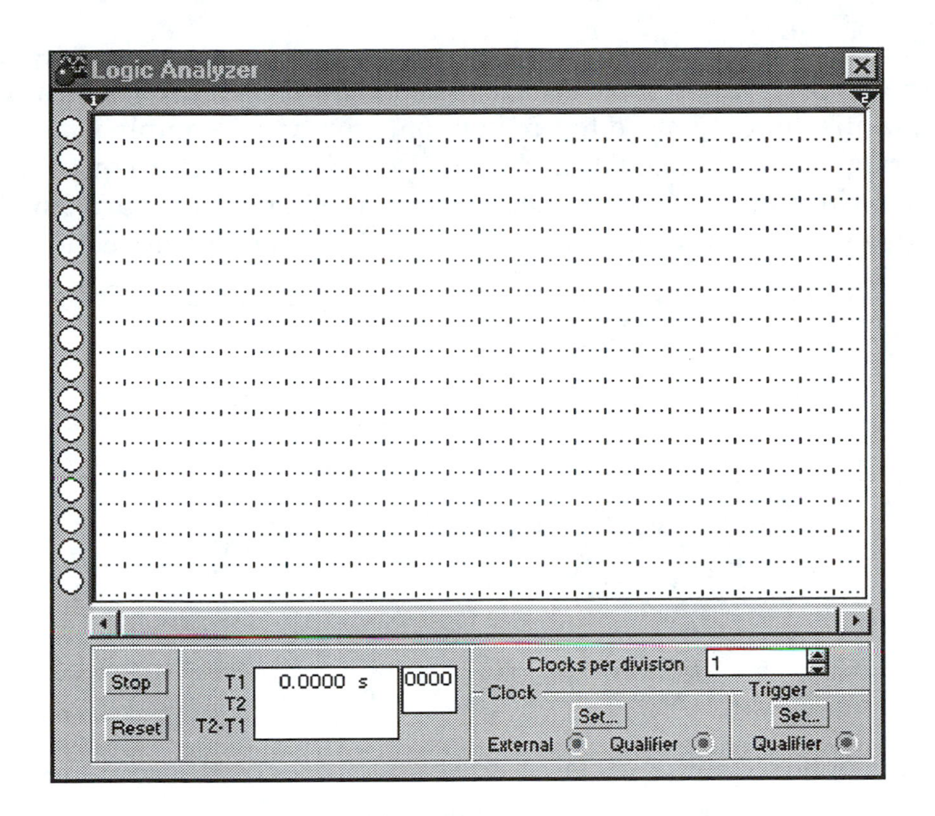

Figure 2.3(b): Logic Analyzer details

The Logic Analyzer captures data at a rate determined by an internal or external clock signal. Both pre- and post-trigger data are captured.

Clicking the Set…button in the clock section brings up the menu shown in Figure 2.4.

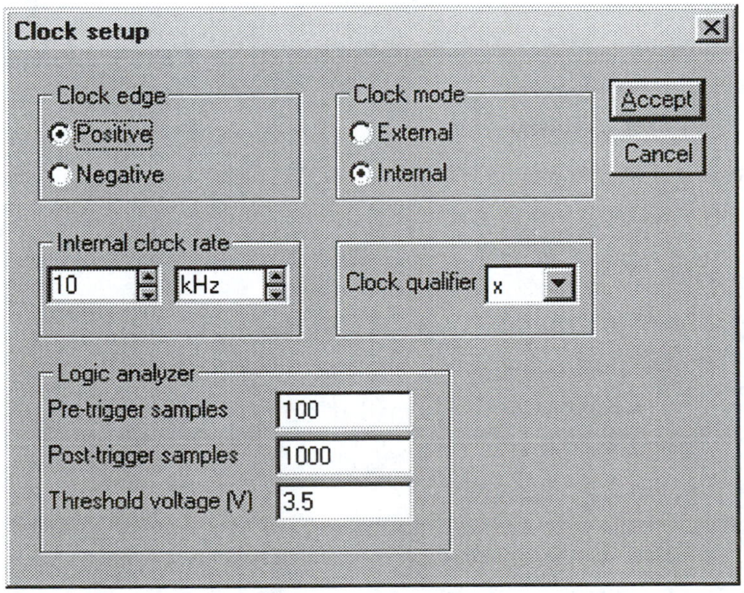

Figure 2.4: Clock setup menu

Note that the number of pre- and post-trigger samples may be specified to tailor the Logic Analyzer to the circuit being investigated. In addition, the Clock qualifier may be used to further control when a sample is taken.

Triggering the Logic Analyzer on a particular pattern or set of patterns is possible through the Trigger patterns menu shown in Figure 2.5. This menu pops up when the Set button is clicked in the Trigger section.

Trigger patterns

A xxxxxxxxxxxxxxxx Accept
B xxxxxxxxxxxxxxxx Cancel
C xxxxxxxxxxxxxxxx
Trigger combinations A
Trigger qualifier 1

Figure 2.5: Trigger patterns menu

Here you can enter a binary pattern to trigger on. In addition, the Trigger combinations option allows several combinations of trigger events to be specified (A OR B, A THEN B, etc.).

Further information on each instrument is available in the Help menu.

Procedure

1. Load the circuit **E2-1**, shown in Figure 2.6.

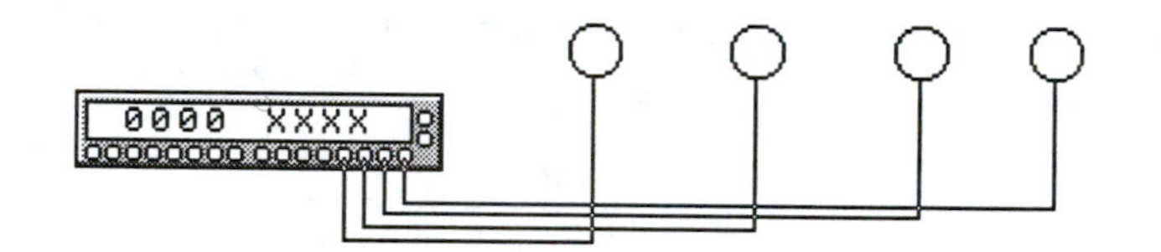

Figure 2.6: Word Generator test circuit

2. Simulate the circuit. What do you see on the logic indicators?

3. Click the Step button. Does the simulation stop? Continue clicking the Step button. What happens?

4. Click the Burst button. What happens?

5. Connect the Logic Analyzer to the four outputs of the Word Generator. Set the clock frequency of the Logic Analyzer to 1 Hz and begin capturing data. What do you see in the Logic Analyzer display?

6. Click the Set button in the Triggering section. Set the last four bits of word A to 0110 (xxxxxxxxxxxx0110) and click Accept. Simulate the circuit again. How does the Logic Analyzer indicate it has been triggered?

7. Use the Word Generator and Logic Analyzer to test the circuit shown in Figure 2.7 (which is found in the file **E2-2**):

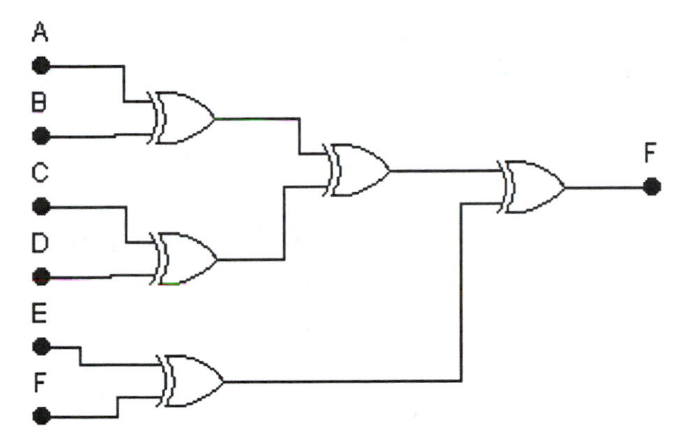

Figure 2.7: Six-input test circuit

The test patterns are shown in Table 2.1.

INPUT	OUTPUT
000000	
100000	
110000	
111000	
111100	
111110	
111111	
101010	

Table 2.1: Test patterns

8. Examine the built-in patterns available with the Word Generator. Save one of them to disk.

9. Load the pattern file **P2-1**. How many samples are stored in it? What are they? Do they have any meaning?

10. Connect a logic indicator to the Data Ready output of the Word Generator. When does it go high during simulation? You may have to adjust the internal clock to obtain a valid display.

11. Use the Data Ready output as a clock input to the Logic Analyzer and experiment with the clock settings until you can capture the Word Generator data.

Discussion

While reviewing your data and results, provide detailed answers to each of the following:

1. What is the value of the Word Generator and the Logic Analyzer?

2. How many triggering options are available in the Logic Analyzer?

3. Does the output column of Table 2.1 represent an odd parity bit or an even parity bit?

4. Explain how the Word Generator can simulate the operation of a RAM or ROM (sequential access only).

Experiment 3

Basic Logic Gates

Name ______________________ **Date** __________

Objectives

The objectives of this experiment are to:

- Examine the operation of each basic logic gate.
- Learn how to determine when a gate is malfunctioning.

Introduction

All digital circuits are composed of combinations of basic logic gates. Even such complex devices as microprocessors are built up from groups of logic gates. In this experiment you will verify the operation of several basic gates and see how truth tables are used to determine if a gate is operating correctly.

Procedure

1. Load the circuit **E3-1**, shown in Figure 3.1.

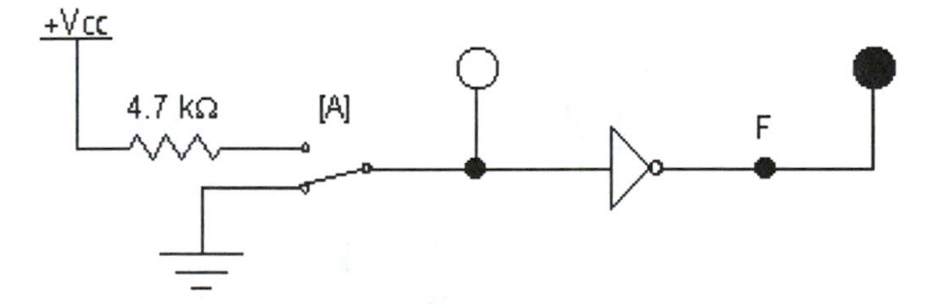

Figure 3.1: Inverter test circuit

2. The switch is used to apply logic zero and logic one values to the inverter. As shown, the switch is applying a logic zero, which causes the inverter to output a logic one, turning on the logic indicator. Pressing the 'A' key causes the switch to change positions. Verify that the circuit is operational by toggling the switch back and forth.

3. Fill in the truth table for the inverter based on the operation of the circuit. Record a zero for A or F if its associated logic indicator is off, and record a one if it is on.

A	F
0	
1	

Table 3.1: Truth table for the inverter

4. Load the circuit file **E3-2**. You should see the circuit shown in Figure 3.2.

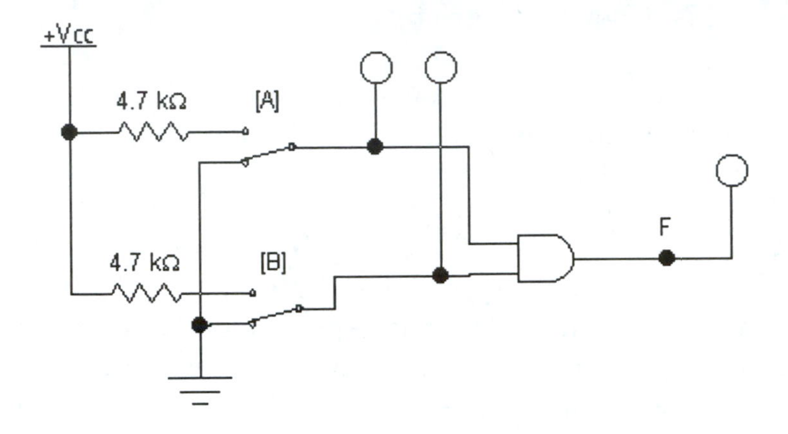

Figure 3.2: AND gate test circuit

5. To test the 2-input AND gate you must apply four different patterns (00, 01, 10, and 11). Verify the operation of the AND gate and fill in its truth table.

A	B	F
0	0	
0	1	
1	0	
1	1	

Table 3.2: Truth table for the AND gate

6. Replace the AND gate with an OR gate (from the Logic Gates parts bin) and complete its truth table.

A	B	F
0	0	
0	1	
1	0	
1	1	

Table 3.3: Truth table for the OR gate

7. Replace the OR gate with a NAND gate and fill in its truth table.

A	B	F
0	0	
0	1	
1	0	
1	1	

Table 3.4: Truth table for the NAND gate

8. Replace the NAND gate with a NOR gate and fill in its truth table.

A	B	F
0	0	
0	1	
1	0	
1	1	

Table 3.5: Truth table for the NOR gate

9. Replace the NOR gate with an exclusive OR gate and fill in its truth table.

A	B	F
0	0	
0	1	
1	0	
1	1	

Table 3.6: Truth table for the exclusive OR gate

10. Replace the exclusive OR gate with an exclusive NOR gate and fill in its truth table.

A	B	F
0	0	
0	1	
1	0	
1	1	

Table 3.7: Truth table for the exclusive NOR gate

11. *Troubleshooting:* Are all of the logic gates in circuit **E3-3** functioning correctly? Figure 3.3 shows the circuit.

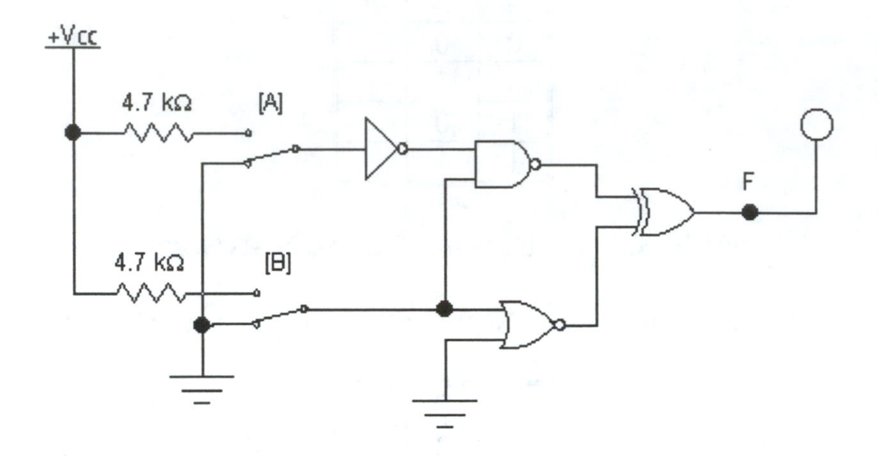

Figure 3.3: Troubleshooting circuit

Discussion

While reviewing your data and results, provide detailed answers to each of the following:

1. What is the relationship between the AND gate and the NAND gate?

2. What would happen if two inverters were connected in series?

3. Which gate outputs a zero only when both inputs are high?

4. Which gate outputs a zero when both inputs are the same logic level?

Experiment 4

DeMorgan's Theorem

Name ______________________ **Date** __________

Objectives

The objectives of this experiment are to:

- Examine the logic conversions possible via DeMorgan's Theorem.
- Convert from one logic form to another using DeMorgan's Theorem.

Introduction

DeMorgan's Theorem provides us with the ability to perform the following logical transformations:

- AND to inverted-input NOR
- NAND to inverted-input OR
- NOR to inverted-input AND
- OR to inverted-input NAND

In this experiment we will verify the equivalence of these transformations.

Procedure

1. Load the circuit **E4-1**, shown in Figure 4.1.

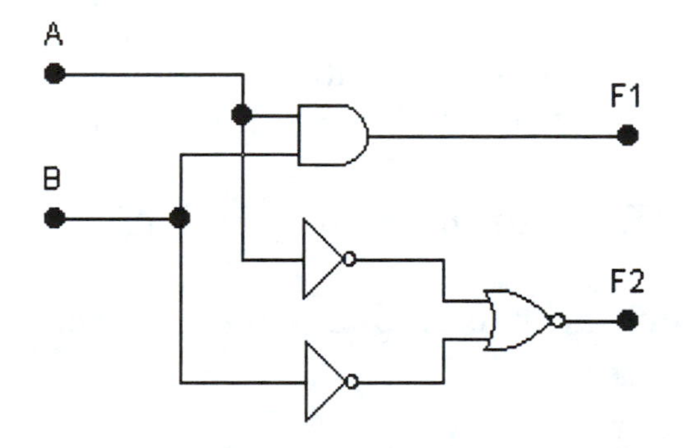

Figure 4.1: AND equals inverted-input NOR

2. Add the necessary components required to test the circuit. Demonstrate that outputs F1 and F2 are always the same.

3. Determine if circuit **E4-2** (Figure 4.2) operates the same as circuit **E4-1**.

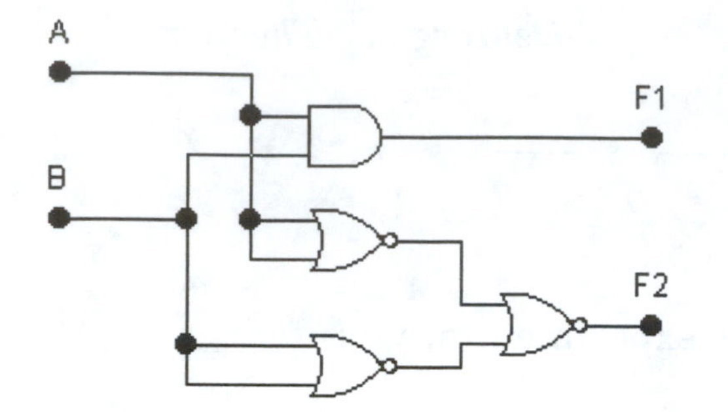

Figure 4.2: Using NOR gates as inverters

The NOR gates that have both inputs tied together operate as inverters.

4. Load circuit **E4-3**, shown in Figure 4.3.

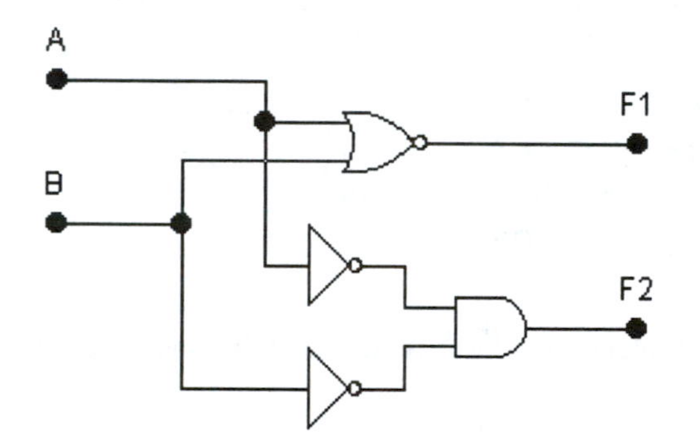

Figure 4.3: NOR equals inverted-input AND

5. Verify that the F1 and F2 outputs are identical for all input combinations.

6. Change the NOR gate to a NAND gate and the AND gate to an OR gate. Demonstrate that NAND equals inverted-input OR.

7. Demonstrate that OR equals inverted-input NAND.

8. What type of logic operation is implemented in circuit **E4-4**, shown in Figure 4.4?

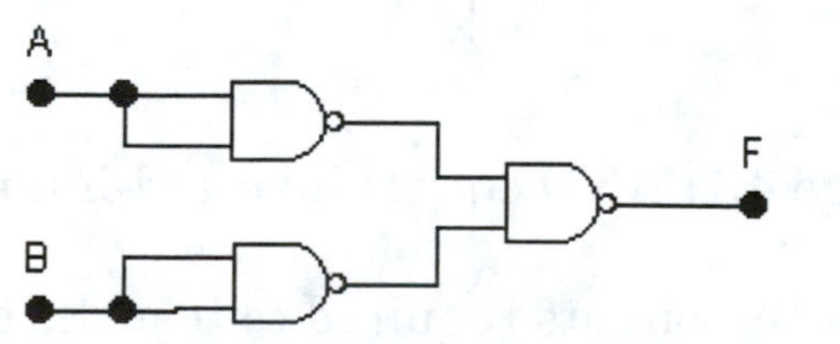

Figure 4.4: All NAND logic

9. Use DeMorgan's Theorem to convert circuit **E4-5** (Figure 4.5) into all NAND logic. Verify that both circuits have the same truth table.

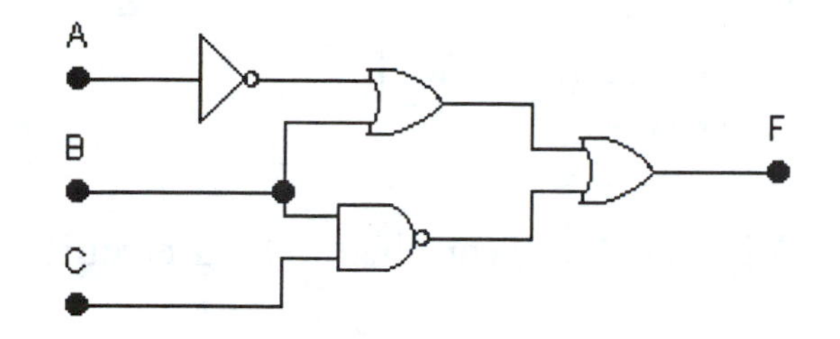

Figure 4.5: Convert into all NAND logic

A	B	C	F
0	0	0	
0	0	1	
0	1	0	
0	1	1	
1	0	0	
1	0	1	
1	1	0	
1	1	1	

Table 4.1: Truth table for the conversion circuit

10. Repeat step 9 for circuit **E4-6**, except convert into all NOR logic. Figure 4.6 shows the original circuit.

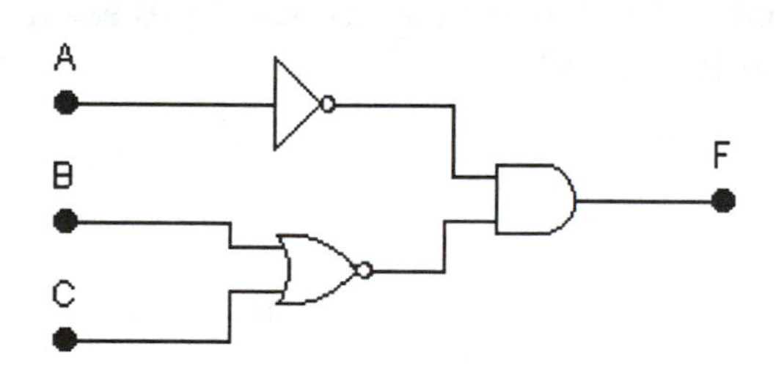

Figure 4.6: Convert into all NOR logic

11. *Troubleshooting:* Does circuit **E4-7** operate like an AND gate? If not, determine why.

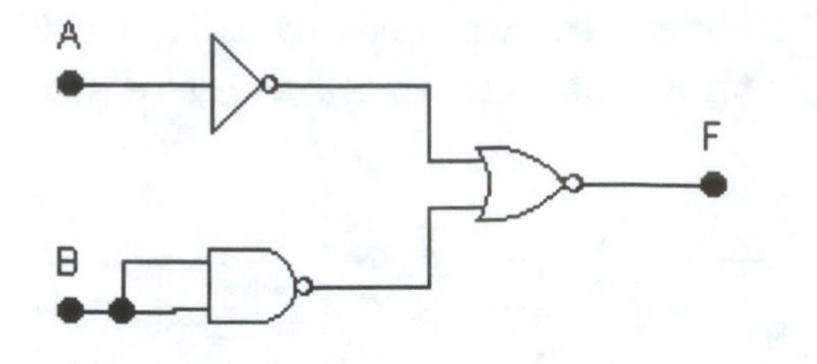

Figure 4.7: Troubleshooting circuit

Discussion

While reviewing your data and results, provide detailed answers to each of the following:

1. Show, using truth tables, that an AND gate is equivalent to an inverted-input NOR gate.

2. Why does a NOR gate with inverted inputs act like an inverter?

3. How is it possible to convert a circuit into all NAND or all NOR logic?

4. If an OR gate is needed but only two NAND gates and an inverter are available, what can be done?

Experiment 5

Combinational Logic Circuits

Name ______________________ **Date** __________

Objectives

The objectives of this experiment are to:

- Examine the relationship between a Boolean expression and a combinational logic circuit.
- Convert Boolean equations into combinational logic circuits.
- Determine the function of a combinational logic circuit.

Introduction

A combinational logic circuit consists of several logic gates connected in such a way that multiple digital inputs are reduced to one or more (in general, fewer) outputs. In this experiment we will examine several combinational logic circuits and their operation.

Procedure

1. Load the circuit **E5-1**, shown in Figure 5.1.

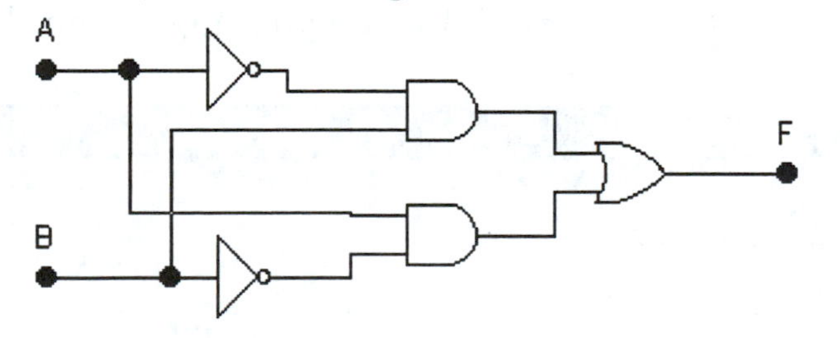

Figure 5.1: Simple combinational logic circuit

2. Determine the truth table for the circuit by simulation.

A	B	F
0	0	
0	1	
1	0	
1	1	

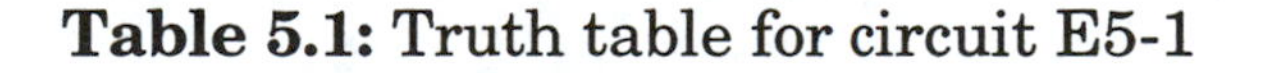

Table 5.1: Truth table for circuit E5-1

3. Open the Logic Converter. When minimized, it looks like Figure 5.2(a). Figure 5.2(b) shows the actual instrument and its controls.

Figure 5.2(a): Logic Converter minimized icon

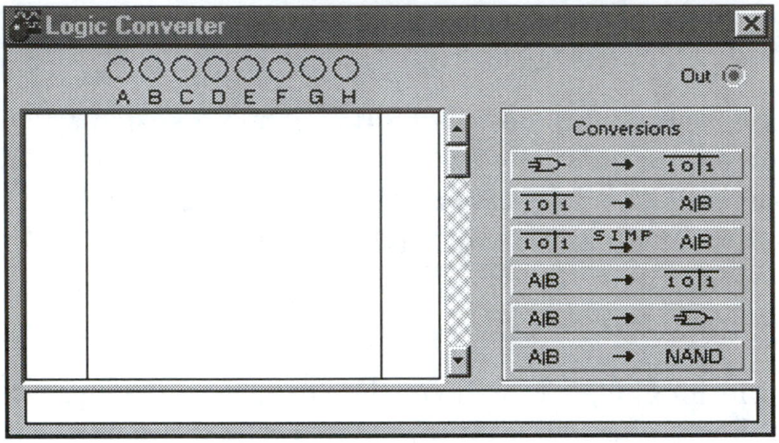

Figure 5.2(b): Logic converter details

The Logic Converter allows you to enter a truth table and generate its associated Boolean expression and/or combinational logic circuit. To begin, left-click inside the circles under A and B at the top of the Logic Converter. Its display should look like Figure 5.3 when you have done this.

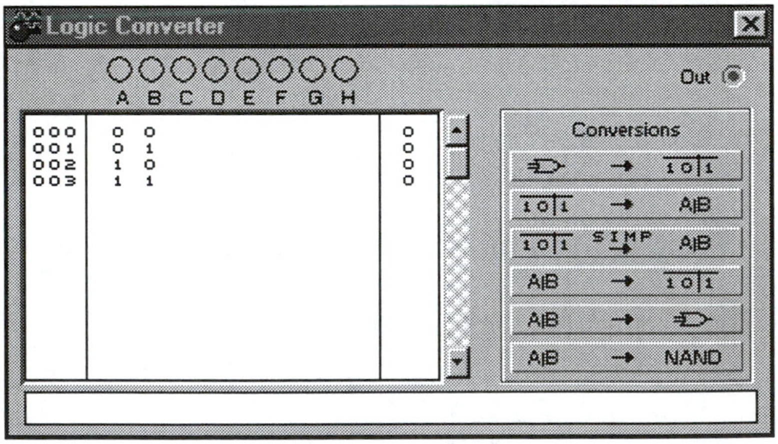

Figure 5.3: Setting up a 2-input truth table

4. Now the 1's in the output column must be added. This is accomplished by left-clicking on the lines of the truth table output where a one is required

and pressing '1'. For our example, enter 1's on the middle two lines of the truth table. Your display should look like Figure 5.4.

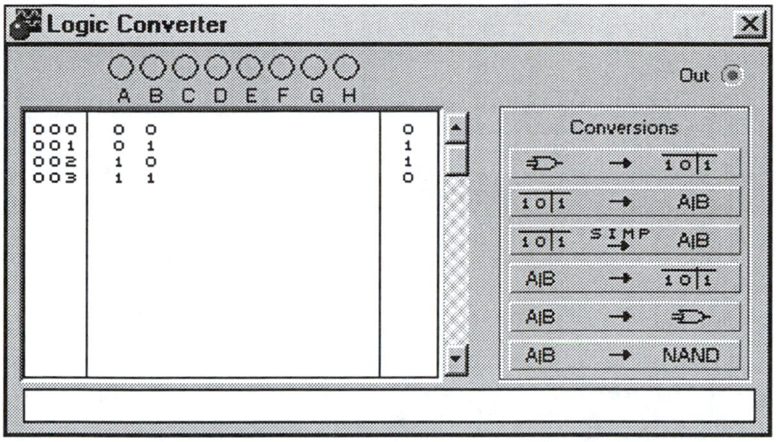

Figure 5.4: Truth table ready for conversion

5. The six conversion buttons on the Logic Converter allow you to convert between equations, truth tables, and logic circuits. Clicking the second button from the top displays the Boolean equation for the truth table in the lower display window, as indicated in Figure 5.5.

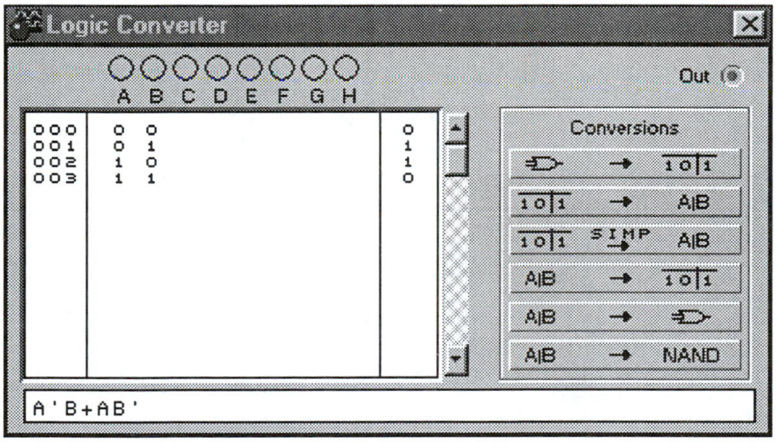

Figure 5.5: Boolean equation for truth table

Notice that the Logic Converter uses a single quote after a variable instead of an overbar to indicate inversion. The equivalent Boolean expression, with overbars, is:

$$F = A\overline{B} + \overline{A}B$$

Which is the expanded version of the simpler expression F = A ⊕ B.

6. Left-clicking the second-to-last button converts the Boolean equation into an actual combinational logic circuit, as indicated in Figure 5.6.

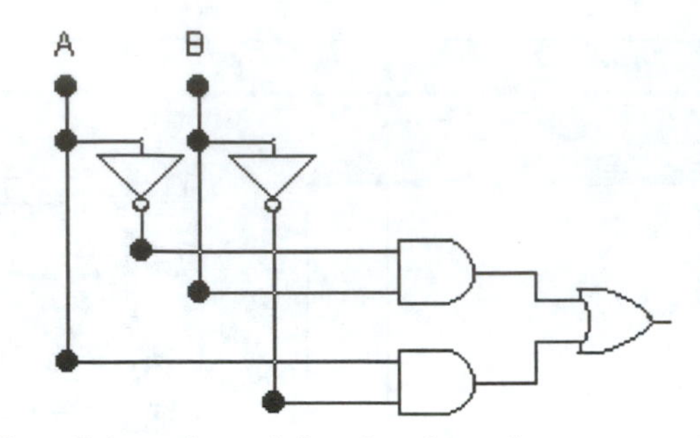

Figure 5.6: Combinational logic circuit generated by equation

7. Repeat steps 4 through 6 for a truth table containing three inputs (A, B, and C). Ones are required on lines 0, 3, 4, and 5.

8. Repeat step 7 except left-click the third button down (containing SIMP) to simplify the Boolean expression. Are there fewer terms? Is less logic required? Does the circuit operate correctly?

9. The Logic Converter is also capable of determining the truth table (or Boolean equation) for a supplied circuit. Open circuit **E5-2**, which indicates the necessary connections (see Figure 5.7).

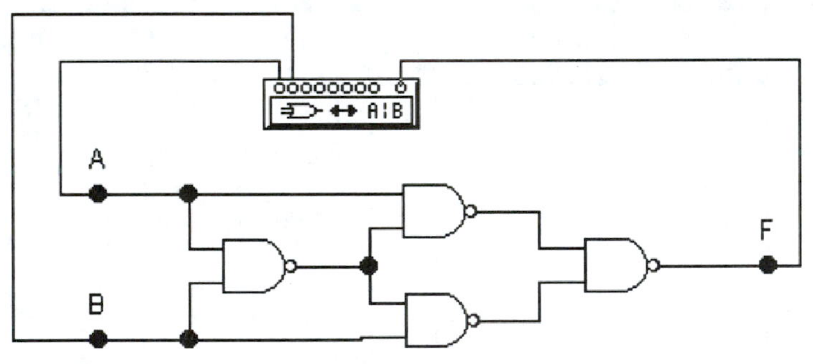

Figure 5.7: Test circuit for Logic Converter

10. Open the Logic Converter and left-click the first conversion button. Record the truth table determined by the Logic Converter.

A	B	F
0	0	
0	1	
1	0	
1	1	

Table 5.2: Truth table for circuit E5-2

11. *Troubleshooting:* Load circuit **E5-3**, shown in Figure 5.8. Connect it to the Logic Converter and determine its truth table. It should match the one shown in Table 5.3. If the output is not correct, determine why.

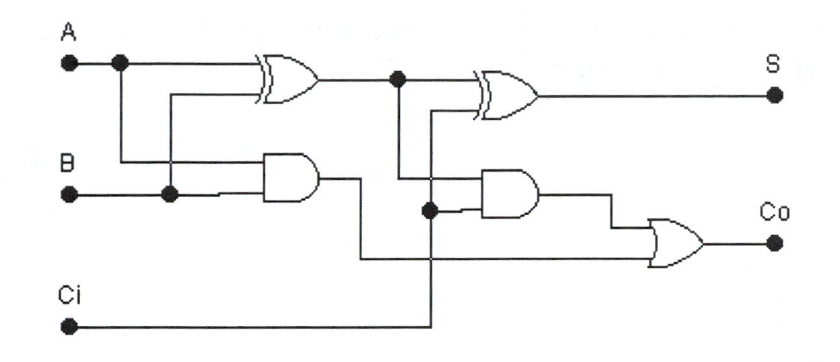

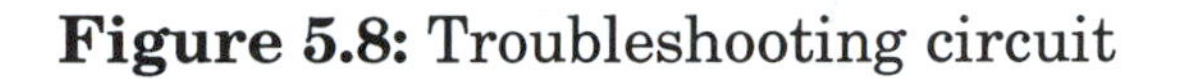

Figure 5.8: Troubleshooting circuit

A	B	CI	S	CO
0	0	0		
0	0	1		
0	1	0		
0	1	1		
1	0	0		
1	0	1		
1	1	0		
1	1	1		

Table 5.3: Truth table for troubleshooting circuit

You must examine one output at a time with the Logic Converter.

Discussion

While reviewing your data and results, provide detailed answers to each of the following:

1. What basic logic function is being performed by the circuit in step 1 (Figure 5.1)?

2. What is the advantage of simplifying a Boolean equation?

3. What basic logic function is being performed by the circuit in step 9?

4. How is the Logic Converter used to determine the Boolean expression for a combinational logic circuit?

Experiment 6

Arithmetic Circuits

Name ______________________ **Date** __________

Objectives

The objectives of this experiment are to:

- Examine how addition and subtraction are performed using logic gates.
- Examine how binary numbers are compared using logic gates.

Introduction

Binary arithmetic is implemented in hardware by direct application of the basic logic gates. In fact, the AND gate performs a simple form of binary multiplication and the OR gate supplies the basic addition property. In this experiment we examine some of the common arithmetic operations performed on groups of bits.

Procedure

1. Load the circuit **E6-1**, shown in Figure 6.1.

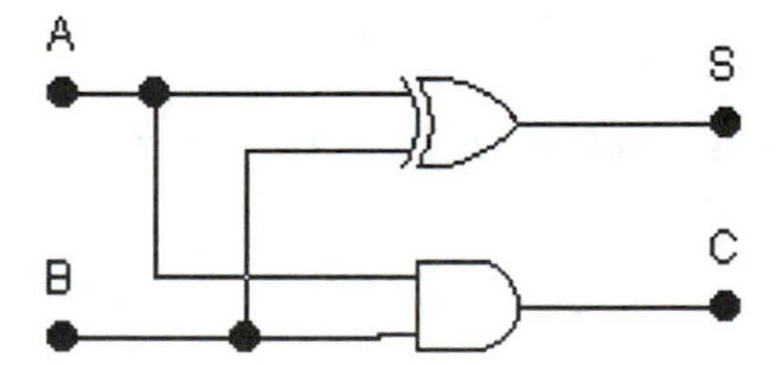

Figure 6.1: Half adder

This circuit is the *half adder*, and is capable of adding two bits together, generating a sum (S) and carry (C). Add the components needed to verify the operation of the half adder, whose truth table is shown in Table 6.1.

A	B	S	C
0	0	0	0
0	1	1	0
1	0	1	0
1	1	0	1

Table 6.1: Truth table for the half adder

2. Circuit **E6-2**, shown in Figure 6.2, performs the same logic function as the half adder. This part is in the Digital parts bin.

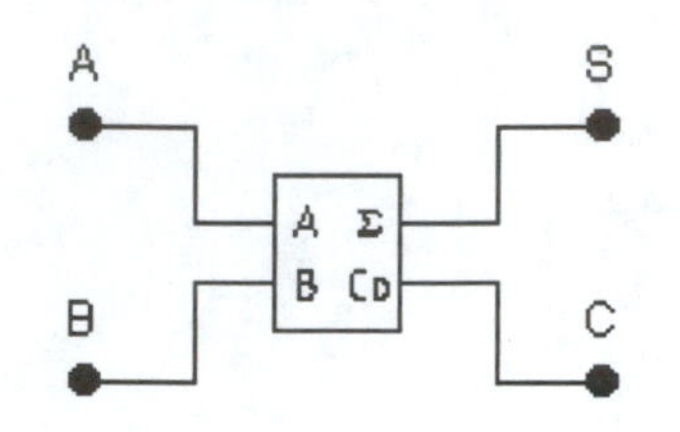

Figure 6.2: Simplified half adder circuit

Test the circuit to verify its operation.

3. A *full adder* adds three bits together. The A and B inputs, as well as a Carry input, are added. Figure 6.3 shows the diagram of the full adder. Load circuit **E6-3** to examine it.

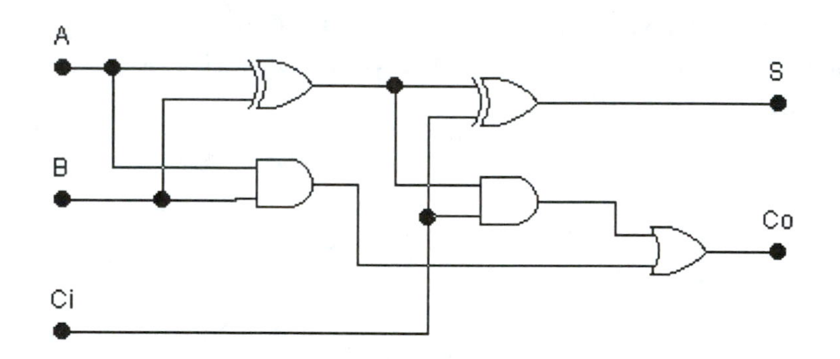

Figure 6.3: Full adder

Use the Logic Converter to test the full adder, whose truth table is shown in Table 6.2.

A	B	CI	S	CO
0	0	0	0	0
0	0	1	1	0
0	1	0	1	0
0	1	1	0	1
1	0	0	1	0
1	0	1	0	1
1	1	0	0	1
1	1	1	1	1

Table 6.2: Truth table for the full adder

4. The Digital parts bin also contains a simplified full adder circuit, which is shown in Figure 6.4. Open **E6-4** to examine it.

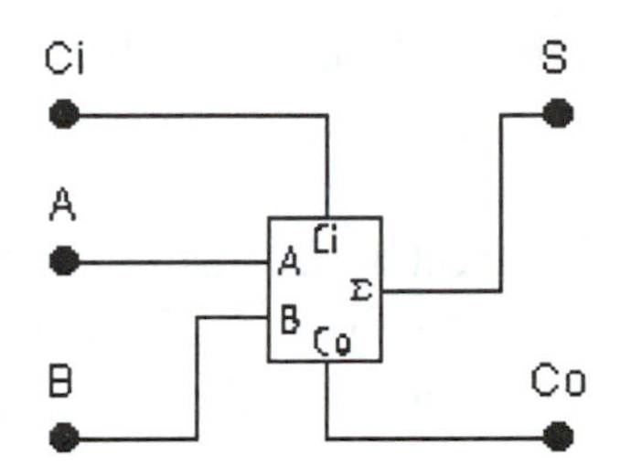

Figure 6.4: Simplified full adder circuit

5. A *parallel* adder adds two or more bits from two input numbers at the same time. Open circuit **E6-5** (shown in Figure 6.5) to examine a 2-bit parallel adder.

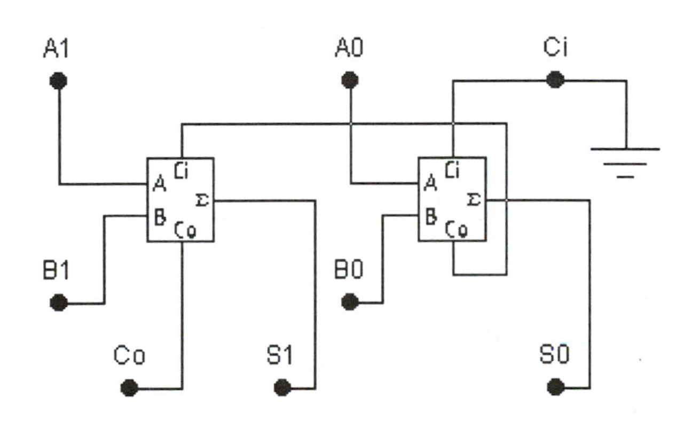

Figure 6.5: 2-bit parallel adder

The Ci input is grounded to apply a logic zero level, guaranteeing that the initial sum bit S0 is correct. Add the components required to test the 2-bit adder.

6. Make a 4-bit parallel adder by cascading two copies of the 2-bit adder. Connect the Co output of the first 2-bit stage to the Ci input of the second 2-bit stage. Note that the sum may now require five bits.

7. Now modify the 4-bit adder into a 4-bit *subtractor*. Subtraction is performed using 2's-complement arithmetic. For example, when subtracting 0011, first complement all bits (1100) and then add one (1100 plus 0001 equals 1101). So, 1101 is the 2's complement of 0011. Our interpretation of both numbers is that 0011 equals positive 3 and 1101 equals negative 3. To subtract, we actually *add* 1101 to the other number. So, in general, A – B becomes A + (–B) and we can perform

subtraction using our adder hardware, with a few additional logic gates. Modify the 4-bit adder by inserting inverters on all the B inputs and by tying the Ci input high. This will simulate the 2's complement of B. Place 0111 on the A inputs and 0011 on the B inputs. The S outputs should equal 0100.

8. It is often necessary to compare two binary numbers to determine if they are equal. Circuit **E6-6** (shown in Figure 6.6) uses exclusive NOR gates to perform the individual comparisons (LSB to LSB, etc.).

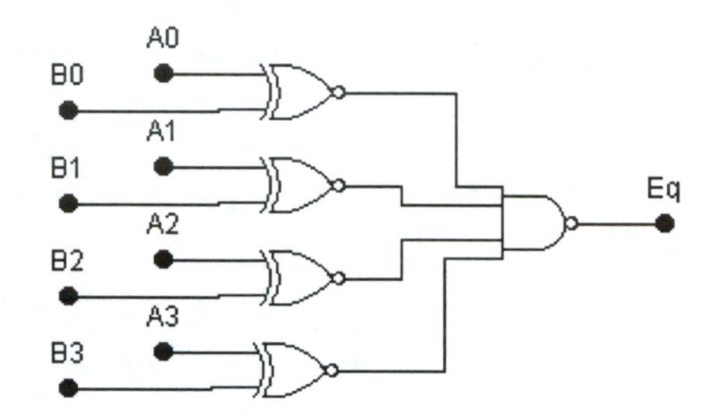

Figure 6.6: 4-bit equality comparator

The Eq output goes low when the 4-bit inputs numbers are equal. Verify the operation of the equality comparator.

9. A *magnitude* comparator is able to determine if two numbers are smaller than, equal to, or larger than one another. A simple 2-bit less-than comparator is shown in Figure 6.7. Open **E6-7** to examine the circuit.

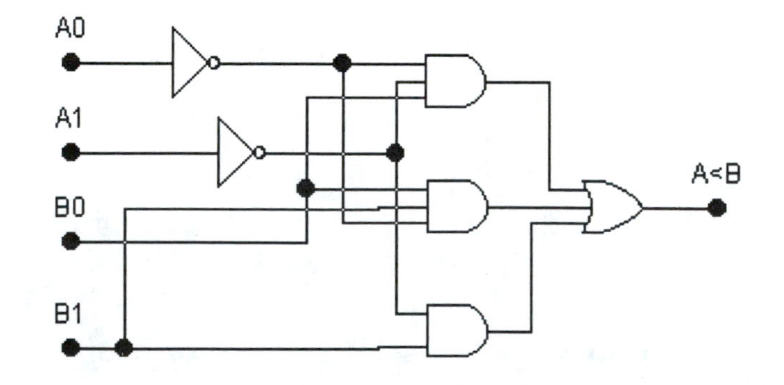

Figure 6.7: 2-bit less-than comparator

A 4-input Karnaugh map was used to design the less-than comparator, with ones placed into each position where A is less than B. Thus, the A<B output goes high whenever A_1A_0 is less than B_1B_0.

Add the required components needed to test the less-than comparator. Fill in the Karnaugh map as you perform the test.

	B_1B_0	B_1B_0	B_1B_0	B_1B_0
A_1A_0	00	01	11	10
00				
01				
11				
10				

Figure 6.8: Karnaugh map for less-than comparator

10. Use two copies of circuit E6-7, plus some additional logic, to make a 4-bit less-than comparator. The additional logic is required because the upper two bits of each number may be equal, forcing the lower 2-bit stage to make the final decision. Thus, the 2-bit less-than comparator must be modified slightly to allow for cascading.

11. *Troubleshooting:* Changing the OR gate on the less-than comparator to a NOR gate makes a greater-than or equal-to comparator. Load circuit **E6-8** (shown in Figure 6.9) and determine if it is operating correctly.

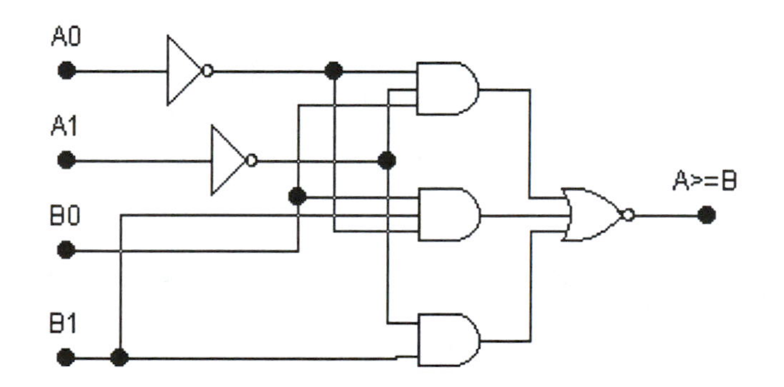

Figure 6.9: Troubleshooting circuit

Discussion

While reviewing your data and results, provide detailed answers to each of the following:

1. What is the difference between a half adder and a full adder?

2. Is it necessary to use a full adder as the first stage in a parallel adder?

3. Why is the Ci input tied high in the 4-bit subtractor?

4. How are exclusive NOR gates used to perform comparisons?

Experiment 7

Multiplexers / Demultiplexers

Name ______________________ **Date** __________

Objectives

The objectives of this experiment are to:

- Examine how multiplexers and demultiplexers are constructed.
- Observe multiplexers and demultiplexers in operation.

Introduction

A multiplexer is a circuit capable of selecting one digital signal from a group of signals and pass it along to its output (similar in operation to a mechanical rotary switch). A demultiplexer performs the opposite function, sending a digital input signal to one of several outputs. These devices provide the ability to transmit several channels of information over a single wire.

Procedure

1. Load the circuit **E7-1**, shown in Figure 7.1.

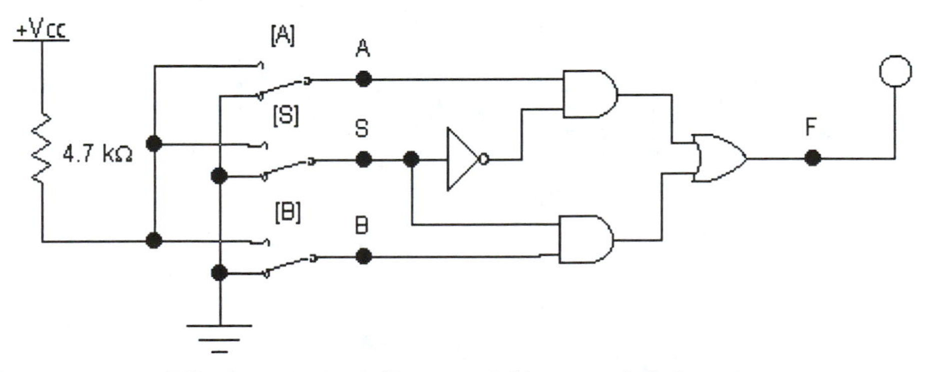

Figure 7.1: 2-line to 1-line multiplexer

2. This circuit is called a 2-line to 1-line multiplexer, or 2:1 multiplexer. The S input selects which of the two inputs A or B is channeled to the output F. Only one AND gate is enabled at a time by the S input, due to the use of the inverter.

   The truth table for the 2:1 multiplexer is shown in Table 7.1. It may be helpful to recall some of the basic rules of Boolean algebra, such as 1•A = A, 0•A = 0, and 0+A = A.

S	F
0	A
1	B

Table 7.1: Truth table for 2:1 multiplexer

Simulate the circuit. Press the 'A' key a number of times. Does the output follow the A input? If not, press 'S' once and try again. Repeat this process to verify that the B signal is channeled to the output. Each time 'S' is pressed the opposite input signal is selected.

3. A 4:1 multiplexer selects one of four signals. Determine how to connect three copies of the 2:1 multiplexer to obtain a 4:1 multiplexer. Use two select inputs (S_1S_0).

4. Circuit **E7-2** shown in Figure 7.2 shows a simpler way to make a 4:1 multiplexer. Add the components required to verify its operation.

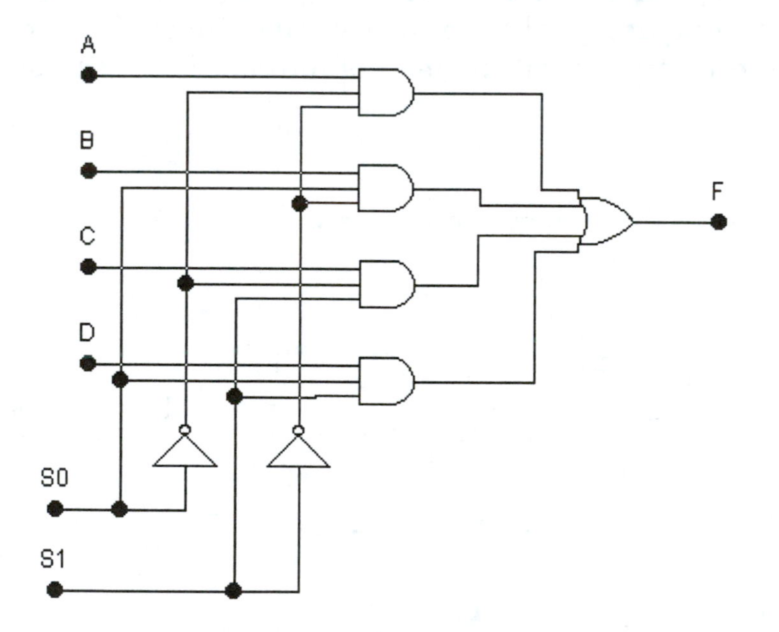

Figure 7.2: 4:1 multiplexer

5. It is often necessary to use more than one multiplexer in a circuit, enabling only one at a time depending on the state of various control or status signals. For example, if the enable input is low, the output stays low regardless of the state of the input and select signals. If the enable input is high, the output follows the selected input. Add the necessary logic to the 4:1 multiplexer so that it utilizes an enable input.

6. Electronics Workbench has a number of built-in 741xx series multiplexer packages. Table 7.2 provides a list of several typical configurations.

PART	FUNCTION
74150	16:1 multiplexer
74151	8:1 multiplexer
74153	Dual 4:1 multiplexer
74157	Quad 2:1 multiplexer

Table 7.2: EWB multiplexer packages

Load the 74151 8:1 multiplexer by locating it in the 741xx Template parts bin. It should look like Figure 7.3.

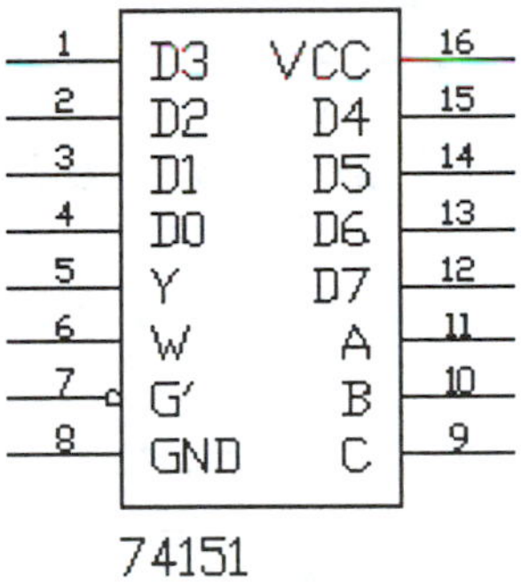

Figure 7.3: 74151 8:1 multiplexer

The A, B, and C inputs accept a three-bit select pattern, which causes the associated data input (D0 through D7) to pass to the Y output. The complement of the signal is available at the W output.

The truth table for the 74151 can be easily viewed by right-clicking on the package and choosing Help from the pop-up menu. The truth table for the 74151 is shown in Figure 7.4.

Add the components required to verify the 74151's operation. You may choose to use the Word Generator to control all of the inputs, or at least provide a pattern to the data inputs, while you use switches to manipulate the three select bits. The Logic Analyzer may also come in handy, displaying all signals (data, select, and output) at the same time for examination.

7. Test the other three multiplexers listed in Table 7.2.

8. Locate other multiplexer packages in the 741xx Template parts bins. How do they differ from those listed in Table 7.2?

Electronics Workbench Help

File Edit Bookmark Options Help

Contents Search Back Print << >>

74151 (1-of-8 Data Sel/MUX)

Select			Strobe	Outputs	
C	B	A	G'	Y	W
X	X	X	1	0	1
0	0	0	0	D0	D0'
0	0	1	0	D1	D1'
0	1	0	0	D2	D2'
0	1	1	0	D3	D3'
1	0	0	0	D4	D4'
1	0	1	0	D5	D5'
1	1	0	0	D6	D6'
1	1	1	0	D7	D7'

Figure 7.4: 74151 truth table

9. Load the circuit **E7-3** (shown in Figure 7.5).

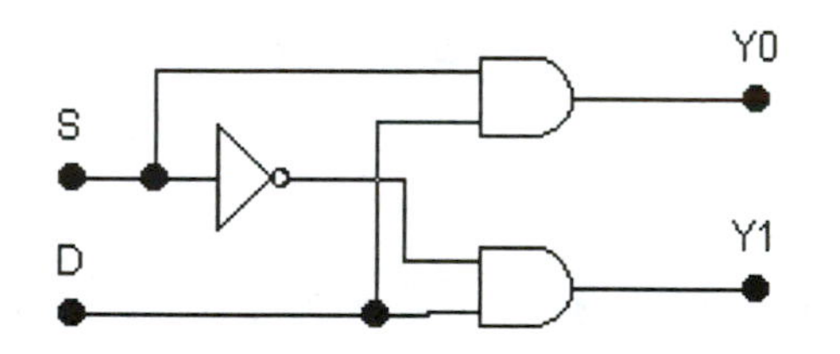

Figure 7.5: 1:2 demultiplexer

The S input determines which output (Y0 or Y1) the data input D is connected to. The truth table for the 1:2 demultiplexer is shown in Table 7.3.

S	Y0	Y1
0	D	0
1	0	D

Table 7.3: Truth table for the 1:2 demultiplexer

Demultiplexers expand one signal into multiple signals, the reverse operation to that of the multiplexer. Verify this behavior by adding the components required to test the demultiplexer.

10. Table 7.4 lists three demultiplexers available in Electronics Workbench. As before, use Help to obtain their truth tables, and add the associated components required to verify the device's operation.

PART	FUNCTION
74139	Dual 2:4 demultiplexer
74154	4:16 demultiplexer
74155	Dual 2:4 demultiplexer

Table 7.4: EWB demultiplexer packages

11. Look for additional demultiplexer packages and note their differences.

12. *Troubleshooting:* Open the circuit **E7-4** shown in Figure 7.6.

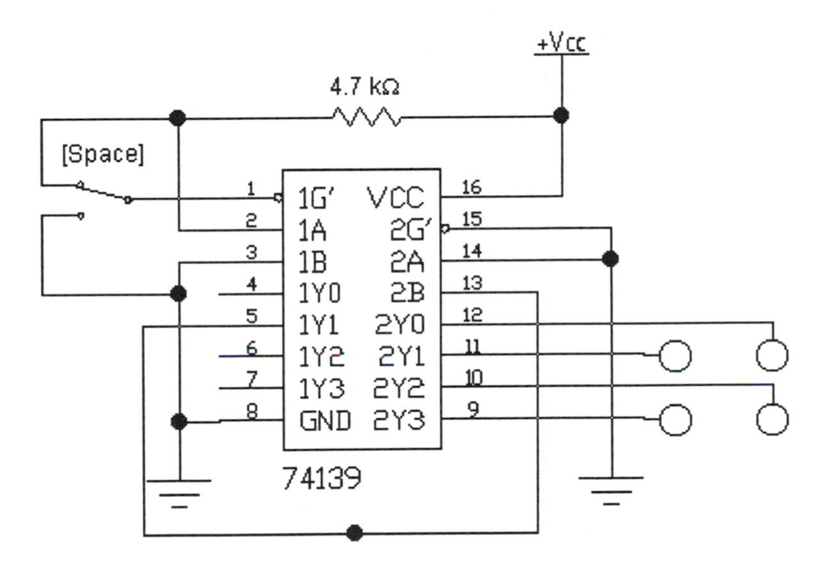

Figure 7.6: Troubleshooting circuit

Predict which logic indicators are on for each position of the switch. Simulate the circuit and verify your predictions.

Discussion

While reviewing your data and results, provide detailed answers to each of the following:

1. Compare the gate counts for the two 4:1 multiplexer circuits in steps 3 and 4.

2. Explain how to make a 64:1 multiplexer out of standard multiplexer packages.

3. What is required to force the unselected outputs of the 1:2 demultiplexer in Figure 7.5 high instead of low?

4. Explain how to make a 1:32 demultiplexer using 1:8 demultiplexers only.

Experiment 8

Encoders / Decoders

Name ______________________ **Date** __________

Objectives

The objectives of this experiment are to:

- Examine the basic operation of encoders and decoders.
- See how a priority encoder operates.

Introduction

Encoders and decoders are similar in operation to the multiplexers and demultiplexers examined in Experiment 7. An encoder uses information available on several inputs to generate a corresponding encoding on its outputs. Typically there are fewer outputs than inputs.

A decoder is used to convert a set of input patterns (such as 4-bit BCD numbers) into a corresponding set of output patterns (such as activating one of ten decimal outputs). In this experiment we will examine the common applications where encoders and decoders are used.

Procedure

1. Load the circuit **E8-1**, shown in Figure 8.1.

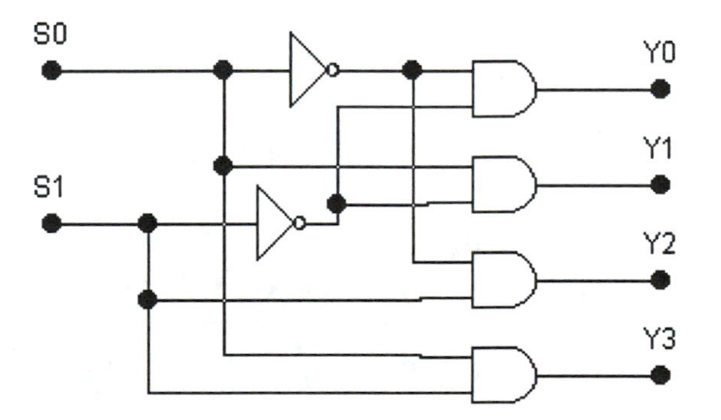

Figure 8.1: One-of-four decoder

2. Add the components required to test the 1:4 decoder. The output selected by the two select inputs goes high. Verify the truth table shown in Table 8.1.

S1	S0	Y0	Y1	Y2	Y3
0	0	1	0	0	0
0	1	0	1	0	0
1	0	0	0	1	0
1	1	0	0	0	1

Table 8.1: Truth table for the 1:4 decoder

3. Show how an enable input (E) can be added to the decoder. A zero on E allows the selected output to go high. All outputs are low when E is high.

4. Load circuit **E8-2**, shown in Figure 8.2.

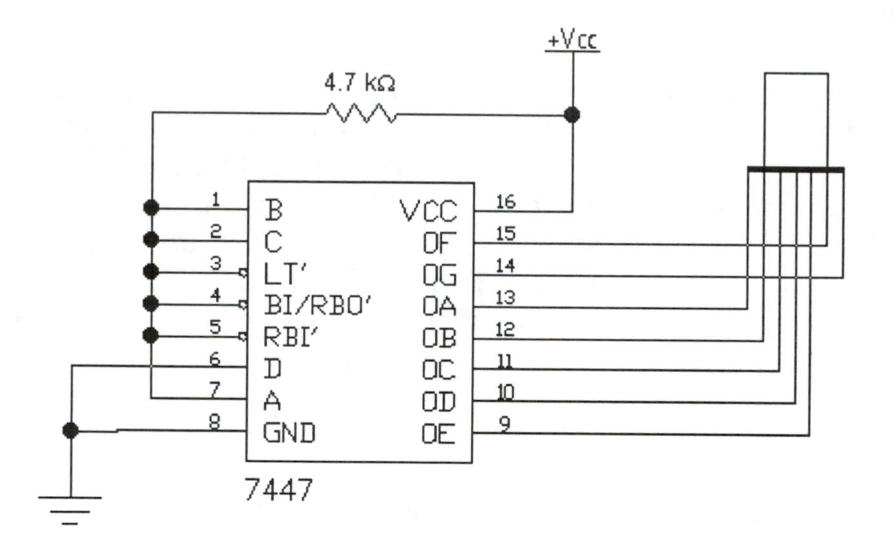

Figure 8.2: 7447 BCD-to-seven-segment decoder

5. Simulate the circuit. What number appears on the display?

6. Modify the circuit as necessary (add 0/1 switches or use the Word Generator) to display the other nine decimal digits.

7. What are the functions of the LT and RBI inputs? Demonstrate their operation.

8. Load the 7445 BCD-to-decimal decoder from the 74xx Template parts bin. It should look like Figure 8.3.

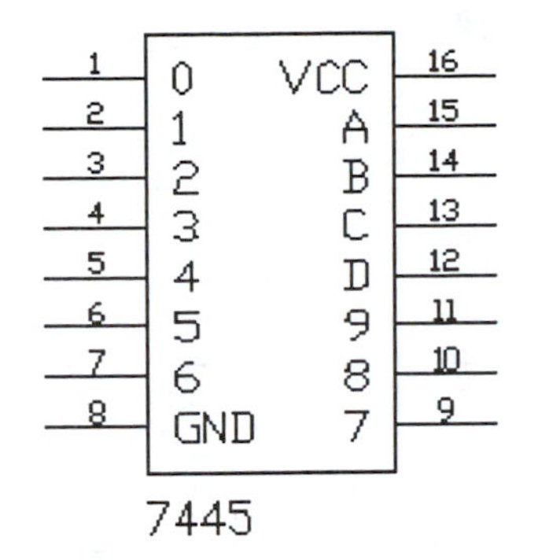

Figure 8.3: 7445 one-of-ten decoder

9. Use the Word Generator to apply input patterns to the A through D inputs (D is the MSB). Examine all inputs and outputs using the Logic Analyzer. Right-click on the 7445 and select Help to view the truth table.

10. Load the 74138 3-line to 8-line decoder from the 741xx Template parts bin. It should look like Figure 8.4.

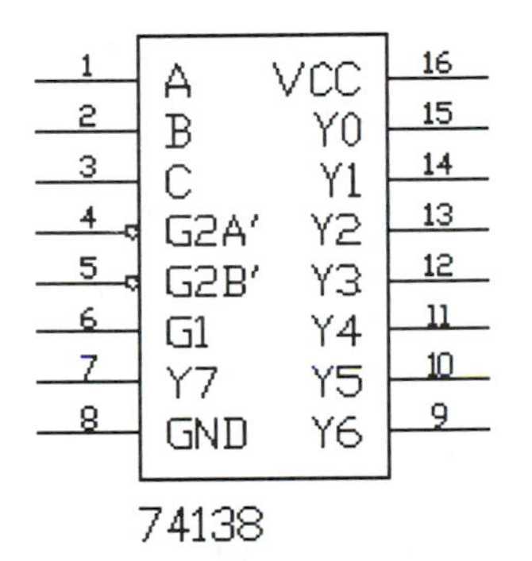

Figure 8.4: 3:8 decoder

11. Add the components required to verify the 74138's operation.

12. Load the circuit **E8-3**, shown in Figure 8.5. This circuit is called a *priority encoder*. The highest numbered input that is active (high) is encoded on the A and B outputs. A simplified truth table is shown in Table 8.2.

3	2	1	0	B	A	S
0	0	0	0	0	0	0
0	0	0	1	0	0	1
0	0	1	0	0	1	1
0	1	0	0	1	0	1
1	0	0	0	1	1	1

Table 8.2: Truth table for the 4:2 priority encoder

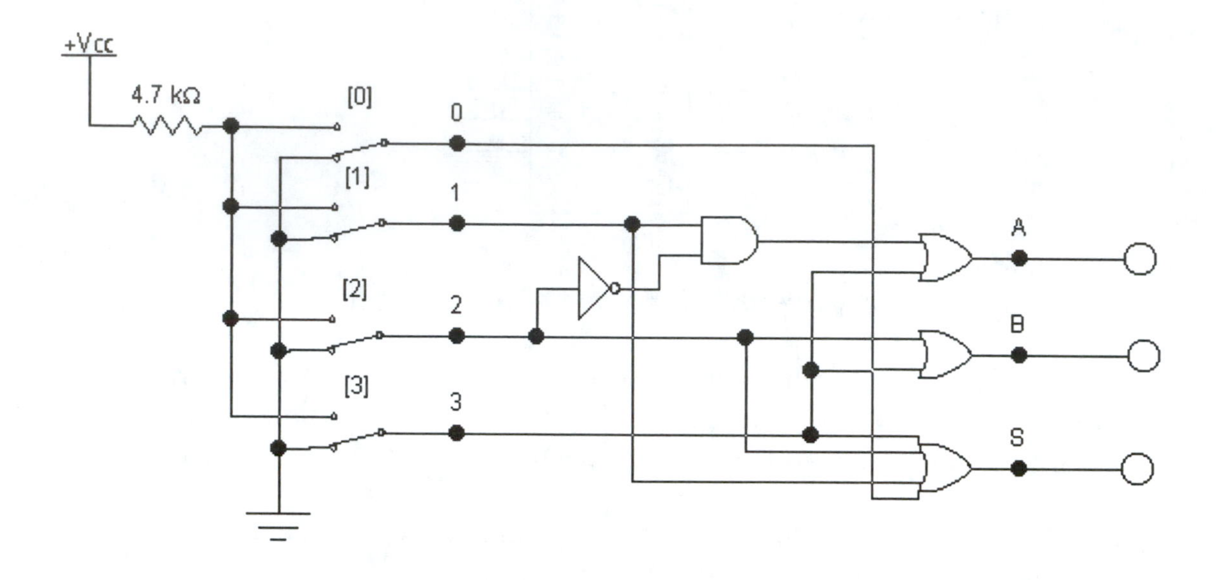

Figure 8.5: 4:2 priority encoder

13. Verify the truth table of the priority encoder.

14. Now apply different combinations of active inputs to the priority encoder (for example, inputs 0 and 2 both high at the same time). How do the outputs reflect a priority?

15. Load the 74148 8-line to 3-line priority encoder from the 741xx Template parts bin. It should look like Figure 8.6.

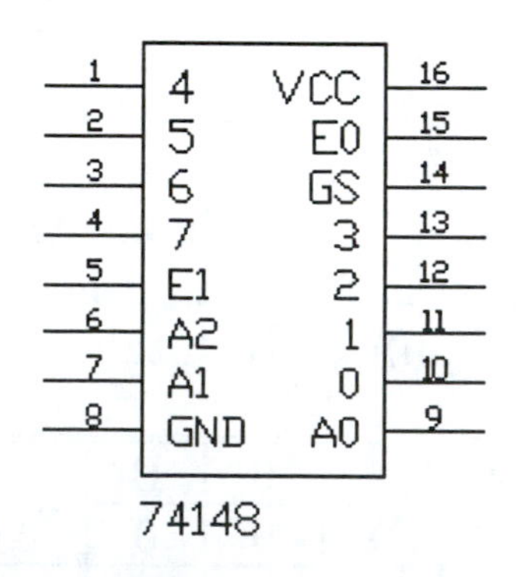

Figure 8.6: 8:3 priority encoder

Add the components required to verify the 74148's truth table.

16. Show how two 74148's can be used to implement a 16:4 priority encoder.

17. Load the 74147 10-line to 4-line priority encoder from the 741xx Template parts bin. It should look like Figure 8.7.

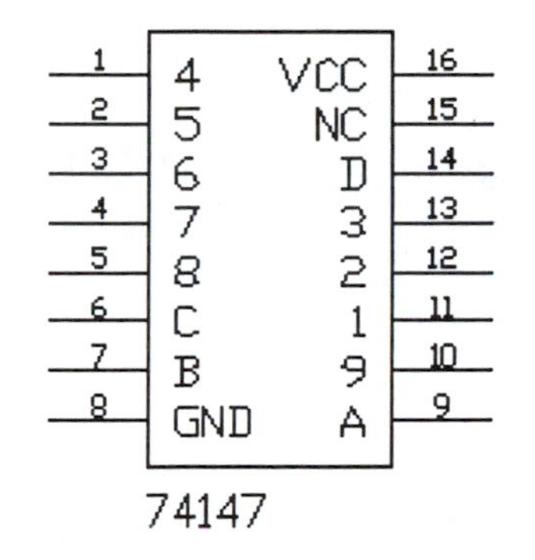

Figure 8.7: 10:4 priority encoder

Add the components required to verify the 74147's truth table.

18. *Troubleshooting:* Load the circuit **E8-4**, shown in Figure 8.8.

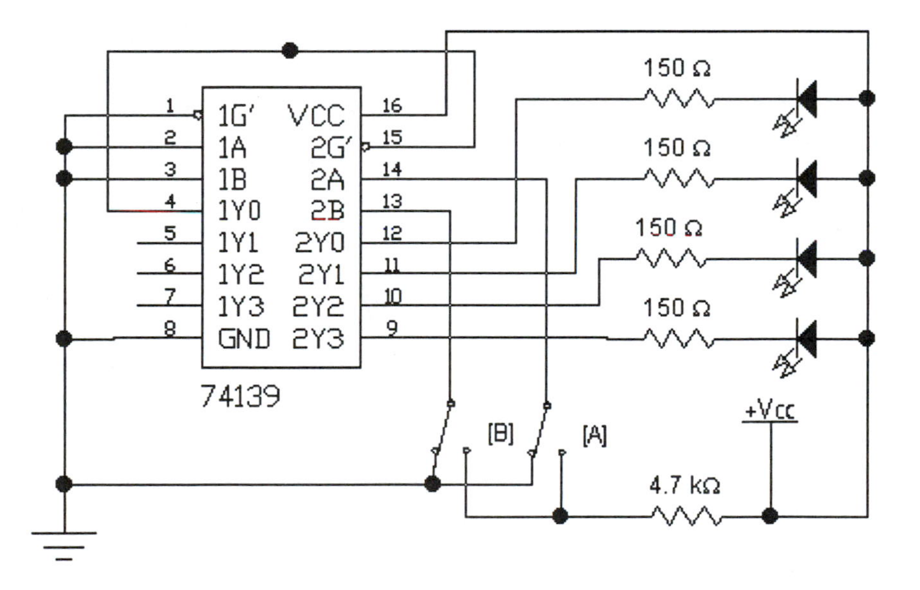

Figure 8.8: Troubleshooting circuit

Is the 2:4 decoder operating correctly?

Discussion

While reviewing your data and results, provide detailed answers to each of the following:

1. How can circuit E8-1 be modified so that the selected output goes low?

2. How do the output patterns of the 7447 differ from those of the 7445?

3. How many 74138's are required to decode 64 outputs? A 6-bit select pattern is required.

4. Based on the truth table of the 4:2 priority encoder, what are the Karnaugh maps for the A and B outputs?

Experiment 9

Digital Oscillators

Name ____________________ **Date** __________

Objectives

The objectives of this experiment are to:

- Examine the operation of the ring oscillator.
- Examine the operation of a crystal oscillator.
- Examine the operation of the 555 timer in astable mode.

Introduction

Many digital circuits require a periodic clock waveform whose rising or falling edges control state changes within the circuit. In this experiment we will examine several ways to generate clock signals.

Procedure

1. Load the circuit **E9-1**, shown in Figure 9.1.

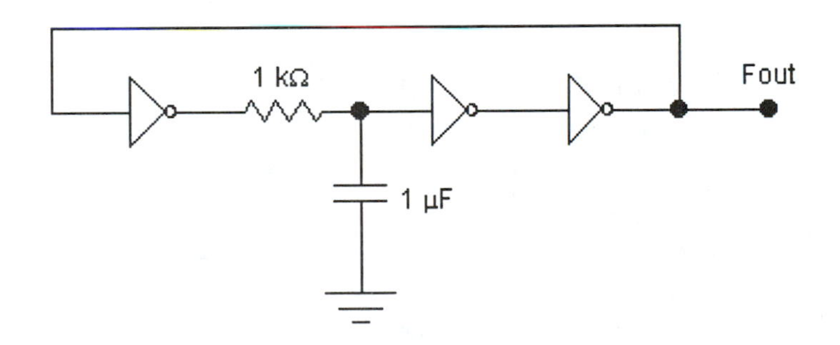

Figure 9.1: Ring oscillator

This circuit is called the *ring oscillator*, due to the circular connection of the inverters. View the output waveform. What is the duty cycle? What is the frequency?

DUTY CYCLE	FREQUENCY

2. Experiment with different values of R and C. Does the frequency change?

3. Replace one of the inverters with a NAND gate. Use the second input on the NAND gate as an enable to turn the ring oscillator on and off.

4. Load the circuit **E9-2**, shown in Figure 9.2.

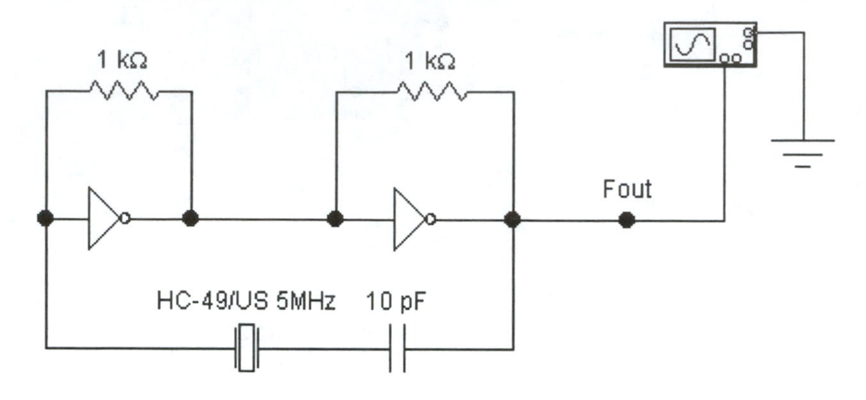

Figure 9.2: Crystal oscillator

When a more stable clock signal is required, a crystal oscillator is preferred over the ring oscillator. Measure the duty cycle and frequency.

DUTY CYCLE	FREQUENCY

Is the frequency correct? Experiment with other crystal types as well.

5. Load the circuit **E9-3**, shown in Figure 9.3.

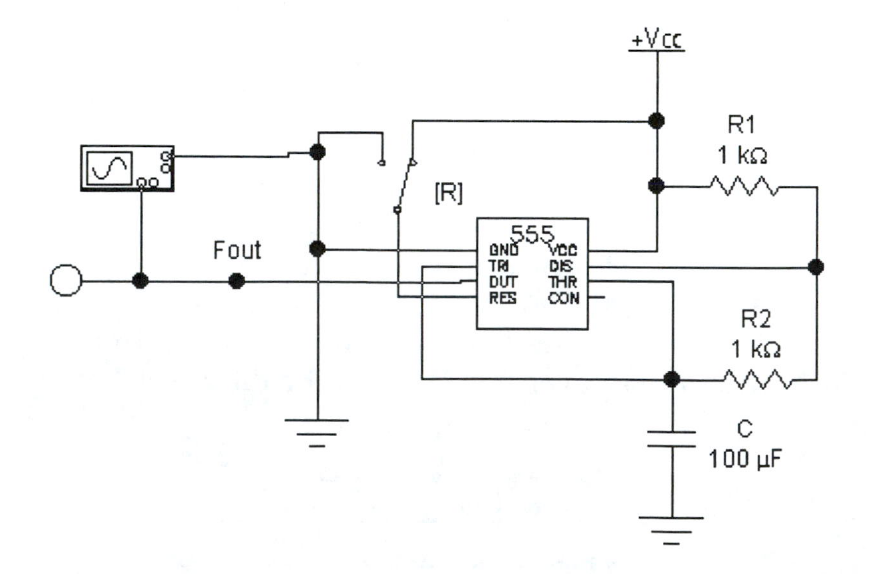

Figure 9.3: 555 timer in astable mode

Simulate the circuit. What does the R switch do?

6. Change the 100 K ohm resistor to 1 K ohm and resimulate the circuit. What are the duty cycle and frequency of operation?

DUTY CYCLE	FREQUENCY

7. How does the frequency from step 6 compare with that predicted by this formula:

$$Fout = \frac{1.43}{(R1 + 2R2)C}$$

8. Change R1 to 4.7 K ohms. What are the duty cycle and frequency?

DUTY CYCLE	FREQUENCY

9. Change R1 back to 1 K ohm and change R2 to 4.7 K ohms. What are the duty cycle and frequency?

DUTY CYCLE	FREQUENCY

10. *Troubleshooting*: Load the circuit **E9-4**, shown in Figure 9.4.

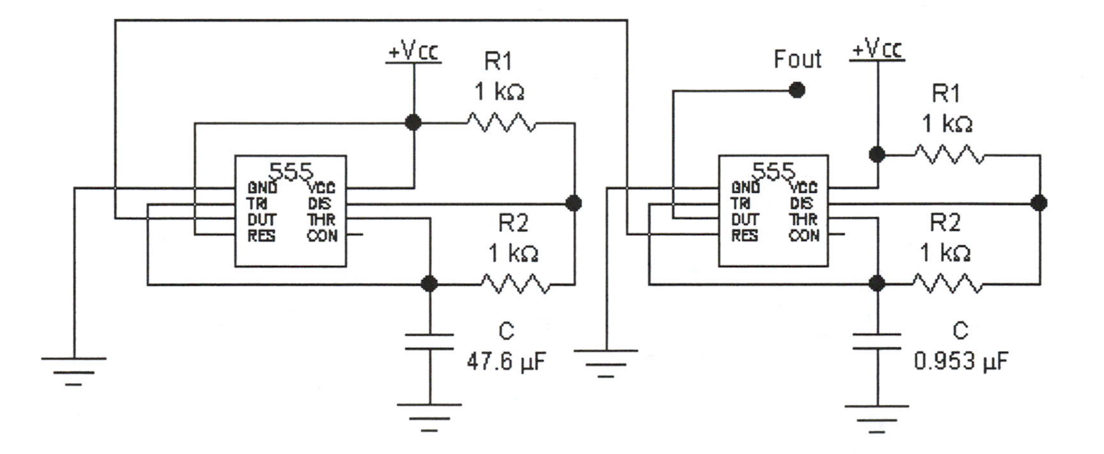

Figure 9.4: Cascaded 555 timers

The first timer should pulse the second timer 10 times per second. The second timer should run at 500 Hz when enabled. Is the circuit operating correctly?

Discussion

While reviewing your data and results, provide detailed answers to each of the following:

1. Is the frequency of the ring oscillator based on the product of R and C?

2. Is the frequency equation for the 555 timer accurate?

3. Determine a formula that predicts the duty cycle for the 555 timer based on the values of R1 and R2.

4. If R1 equals 2.2 K ohms, what values of R2 and C are required to obtain an 80% duty cycle and a frequency of 100 Hz?

Experiment 10

Flip Flops

Name ______________________ **Date** __________

Objectives

The objectives of this experiment are to:

- Examine the operation of the RS latch.
- Examine the operation of the D-type flip flop.
- Examine the operation of the JK flip flop.

Introduction

A flip flop is an essential building block of many important and useful circuits, such as counters, shift registers, and memories. Basically, a flip flop stores a zero or a one. One or more inputs are provided to change the state of the flip flop. In this experiment we will examine the operation of several common flip flops.

Procedure

1. Load the circuit **E10-1**, shown in Figure 10.1.

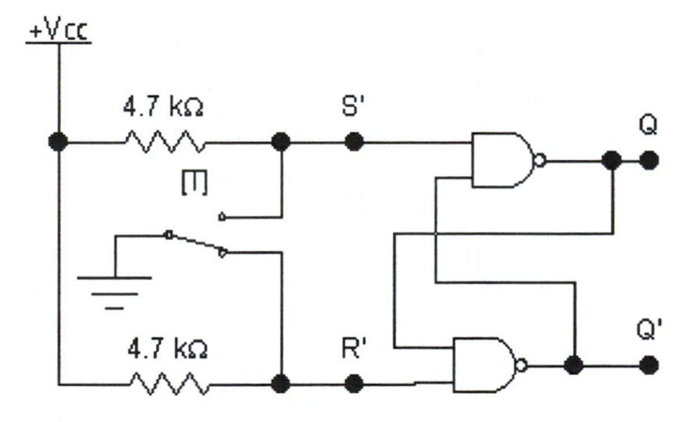

Figure 10.1: RS latch

Attach logic indicators and simulate the circuit. Which indicator is on?

2. Press 'T' several times. What happens at the Q and Q′ outputs?

3. In the real world, the SPDT switch suffers from mechanical contact bounce, which in turn generates hundreds or even thousands of edges

before settling. The RS latch debounces the switch, providing one clean edge on its output for each toggling of the input. This edge can be used to provide a single clock pulse to a flip, counter, or shift register. Electronics Workbench contains a built-in RS latch in the Digital parts bin. Load the circuit **E10-2**, shown in Figure 10.2, to examine its operation.

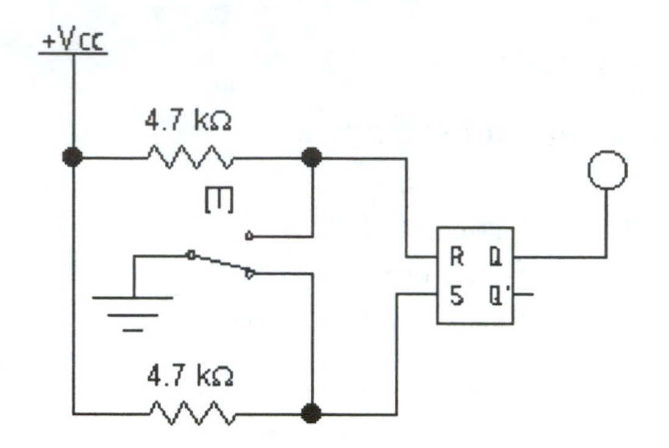

Figure 10.2: Built-in RS latch

4. Load the circuit **E10-3**, shown in Figure 10.3, which uses the RS latch as the clock-pulse circuit for a D-type flip flop. The data input to the flip flop is controlled by the D switch.

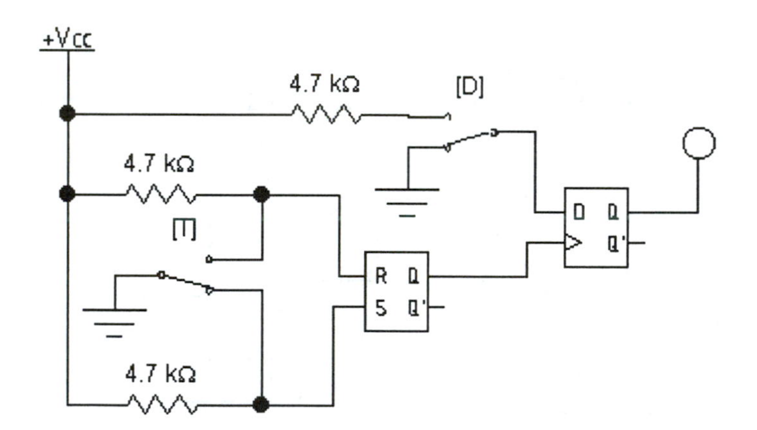

Figure 10.3: Type-D flip flop

5. Determine what edge the D flip flop responds to (positive or negative).

6. Load the circuit **E10-4**, shown in Figure 10.4. The D flip flop has its Q′ output wired back to the D input. This enables a special mode called *toggle*, where the state of the output alternates between zero and one every two clock cycles. Verify this by pressing 'T' ten times. How many times does the logic indicator flash?

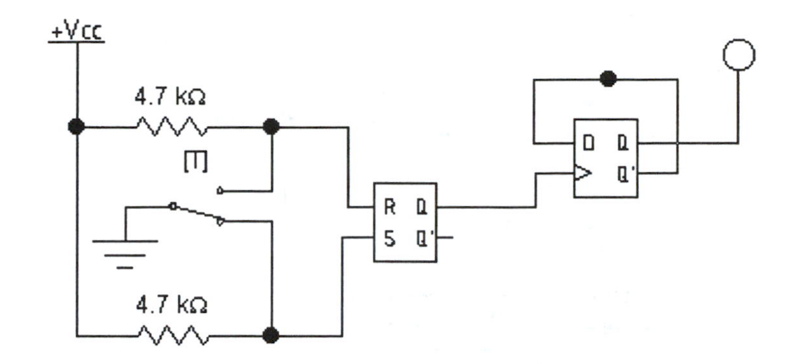

Figure 10.4: D flip flop wired for toggle mode

7. Load the circuit **E10-5**, shown in Figure 10.5.

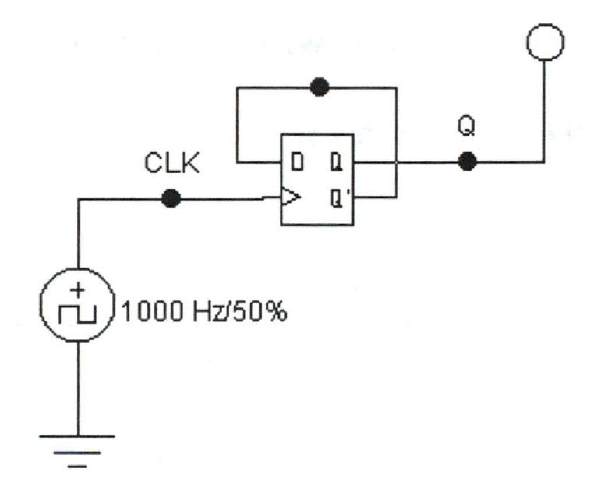

Figure 10.5: Divide-by-two circuit

Verify (using the oscilloscope or logic analyzer) that the frequency of the waveform at the Q output is 500 Hz.

8. Add a second D flip flop (wired for toggle). The Q output of the first flip flop connects to the clock input of the second. What is the frequency at the second Q output?

9. Load the circuit **E10-6**, shown in Figure 10.6. A new D flip flop containing *asynchronous* preset and clear inputs is used. The upper input (connected to the P switch) is the Preset input, which causes Q to go high, regardless of the clock state, whenever Preset is low. The lower input is the active-low Clear input, which forces Q low when active. Verify the operation of the Preset and Clear inputs by toggling 'P' and 'C' during simulation.

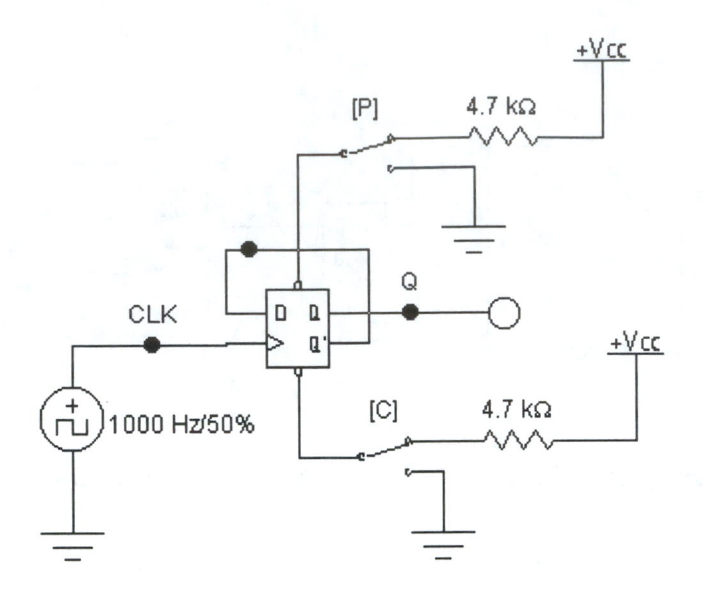

Figure 10.6: D flip flop with asynchronous inputs

10. Load the circuit **E10-7**, shown in Figure 10.7.

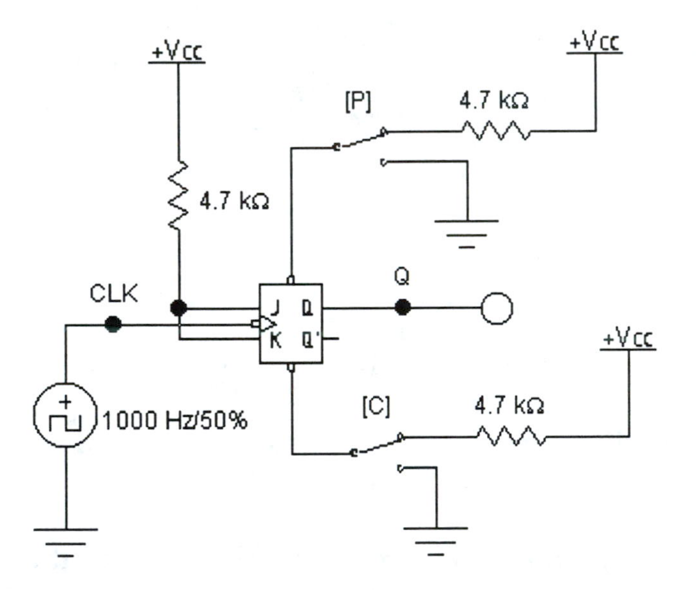

Figure 10.7: JK flip flop wired for toggle mode

11. With the J and K inputs tied high, the output of the JK flip flop will toggle to the opposite state upon each negative edge of the clock. Verify that the JK flip flop is negative-edge triggered.

12. Verify that the Preset and Clear inputs are active low (using 'P' and 'C').

13. The truth table for the JK flip flop is shown in Table 10.1. Verify the four modes of operation.

J	K	Q	Q′
0	0	N/C	N/C
0	1	0	1
1	0	1	0
1	1	Q′	Q

Table 10.1: Truth table for the JK flip flop

When J and K are both low, the outputs do not change state when the flip flop is clocked. When J is low and K is high, Q goes low after the next clock pulse. This mode is called *clear*. When J is high and K is low, Q is *set* after the next clock pulse. Toggle mode is selected when J and K are both high.

14. *Troubleshooting:* Load the circuit **E10-8**, shown in Figure 10.8.

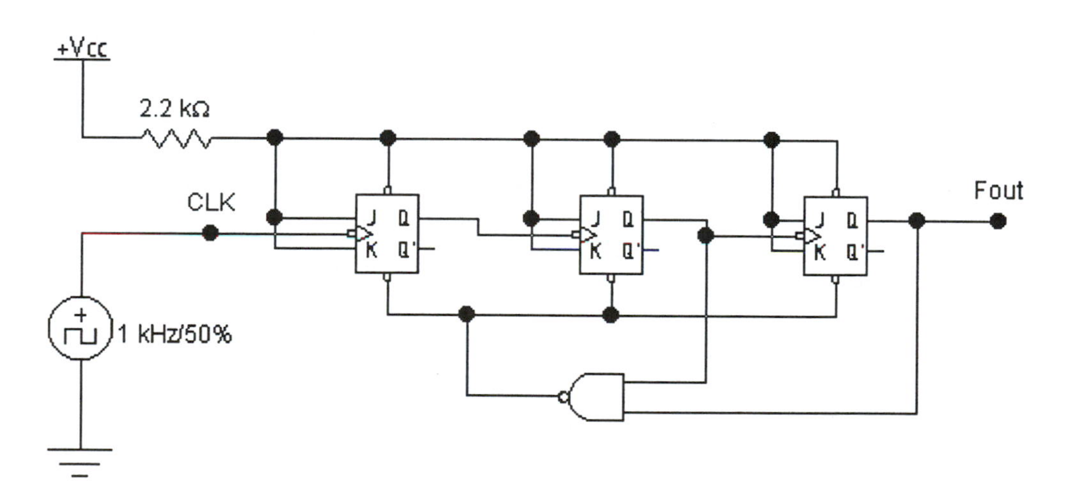

Figure 10.8: Troubleshooting circuit

The frequency at Fout should equal one-sixth the CLK frequency. Determine if the circuit is operating properly.

Discussion

While reviewing your data and results, provide detailed answers to each of the following:

1. How does the built-in RS latch in the Digital parts bin differ from the NAND gate circuit in E10-1?

2. Do the type-D and JK flip flops respond to the same clock edge?

3. Explain how toggle mode is the same as division by two.

4. What is the difference between a synchronous input (D, J, or K) and an asynchronous input (Preset or Clear)?

Experiment 11

The Digital Combination Lock

Name ______________________ **Date** __________

Objective

The objective of this experiment is to:

- Examine the operation of a digital combination lock.

Introduction

In this experiment, a digital combination lock is created out of flip flops. The order in which the flip flops are set is controlled by specific numeric keys. All flip flops must be set to *unlock* the circuit (set the output high).

Procedure

1. Load the circuit **E11-1**, shown in Figure 11.1.

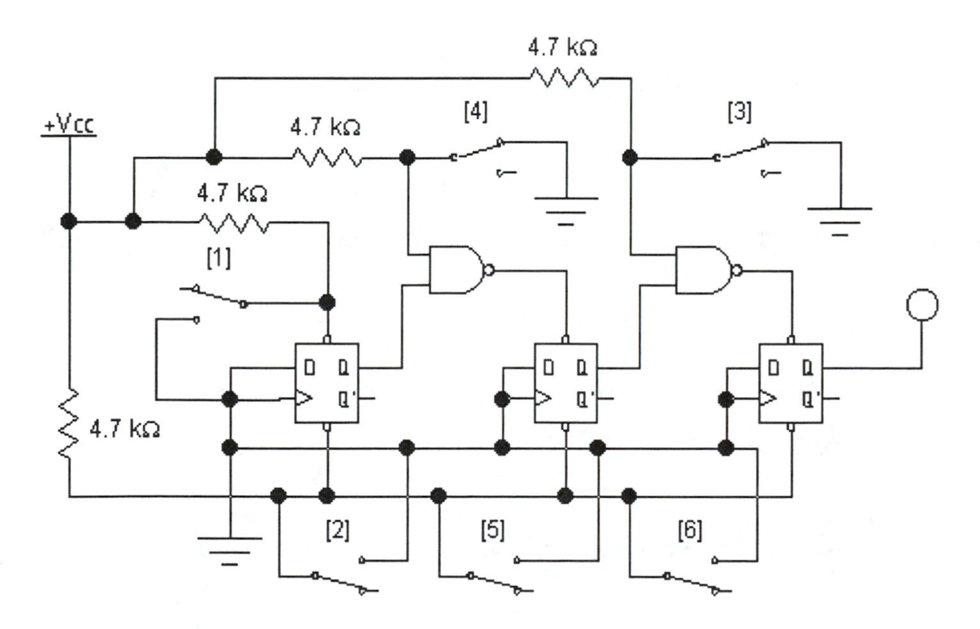

Figure 11.1: Digital combination lock

2. Simulate the circuit. The logic indicator at the output should be off. If not, press '2', '5', or '6' to reset the circuit. Operate the switches as if they were

momentary contact switches, returning them to their original positions (by pressing them twice).

3. Now enter, in sequence, '1', '4', and '3'. Does the output go high?

4. Press any of the other three switches ('2', '5', or '6'). Does the output go low?

5. Is it possible to unlock the circuit using a different sequence, such as '3', '4', and '1'? Remember to treat the switches as momentary contact pushbuttons.

6. Extend the combination lock to five digits.

7. Redesign the combination lock using JK flip flops.

8. *Troubleshooting:* Load the circuit **E11-2**, which is identical to E11-1 except for a hidden fault. Determine the nature of the fault.

Discussion

While reviewing your data and results, provide detailed answers to each of the following:

1. What are the NAND gates used for in circuit E11-1?

2. Why is it OK to ground the clock and data inputs in circuit E11-1?

3. Why is it important to use momentary contact switches?

4. What advantage, if any, does the JK combination lock provide?

Experiment 12

Memory Circuitry

Name ____________________ **Date** __________

Objectives

The objectives of this experiment are to:

- Examine the operation of a random access memory.
- Learn how decoders/multiplexers are used during read/write operations.
- Examine the operation of a simple ROM.

Introduction

Electronics Workbench does not contain a RAM element. Even so, we can study the operation of a simple RAM composed of the same functional blocks used in an actual RAM, although at a smaller scale. In this experiment we examine the operation of a 4-bit RAM composed of SR latches. Two address lines are used to select a bit to read or write.

The operation of a diode-matrix ROM is also examined, to explore the nature of read-only circuitry.

Procedure

1. Load the circuit **E12-1**, shown in Figure 12.1.

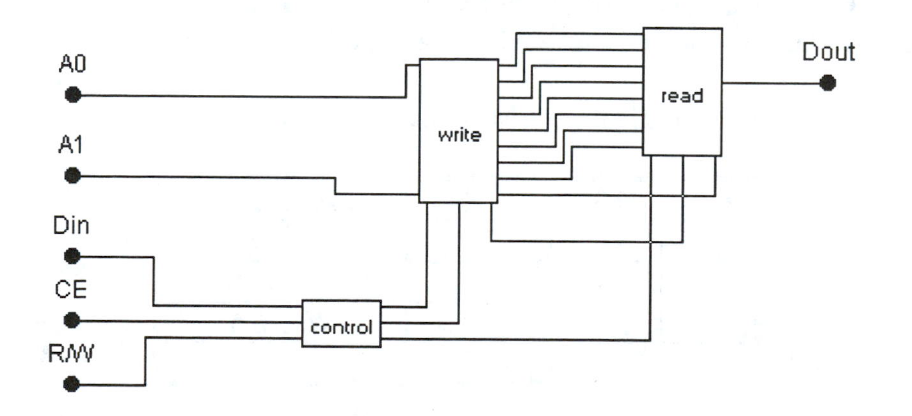

Figure 12.1: Simple 4-bit RAM

Subcircuits are used to help reduce the complexity of the schematic. The *control* subcircuit uses the control inputs CE and R/W to generate the appropriate internal control signals. The *write* subcircuit generates the signals required to set/clear the addressed storage element. The *read* subcircuit contains a 4-bit storage array and output multiplexer.

2. Double-click the control subcircuit to expand it and view its contents. It should look like Figure 12.2.

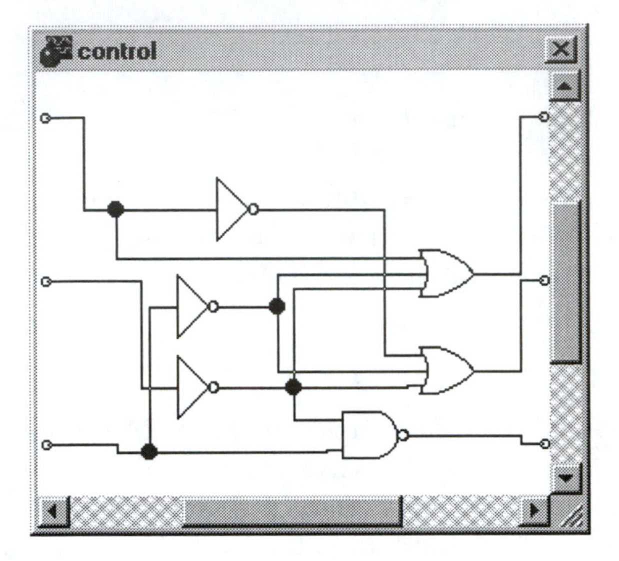

Figure 12.2: Control subcircuit

Din is the data input, R/W is the read/write input, and CE is the chip enable input. R/W must be high during a read operation, and low during a write operation. The CE input must be low to read or write.

- When does the output of the upper OR gate go low?
- When does the output of the lower OR gate go low?
- When does the output of the NAND gate go low?

3. Verify the truth table of the RAM, shown in Table 12.1.

CE	R/W	DIN	OPERATION
1	X	X	None
0	0	0	Write 0
0	0	1	Write 1
0	1	X	Read

Table 12.1: 4-bit RAM truth table

4. Double-click the read subcircuit. It should look like Figure 12.3.

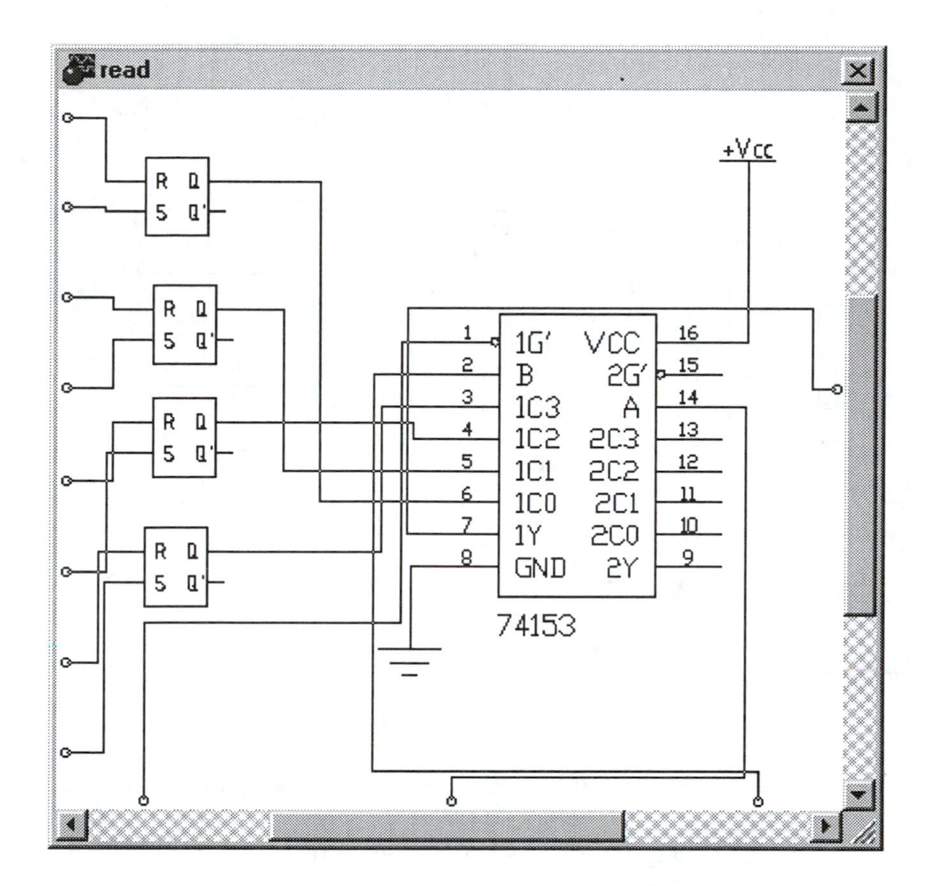

Figure 12.3: Read subcircuit

Each SR latch stores a single bit. The SR inputs from the write subcircuit are used to set/clear a specific bit. The 74153 selects one of the four bits as the output data.

5. Verify the operation of the read subcircuit by storing and reading back each of the bit patterns in Table 12.2.

0001
0010
0100
1000

Table 12.2: Test patterns

6. Double-click the write subcircuit. It should look like Figure 12.4.

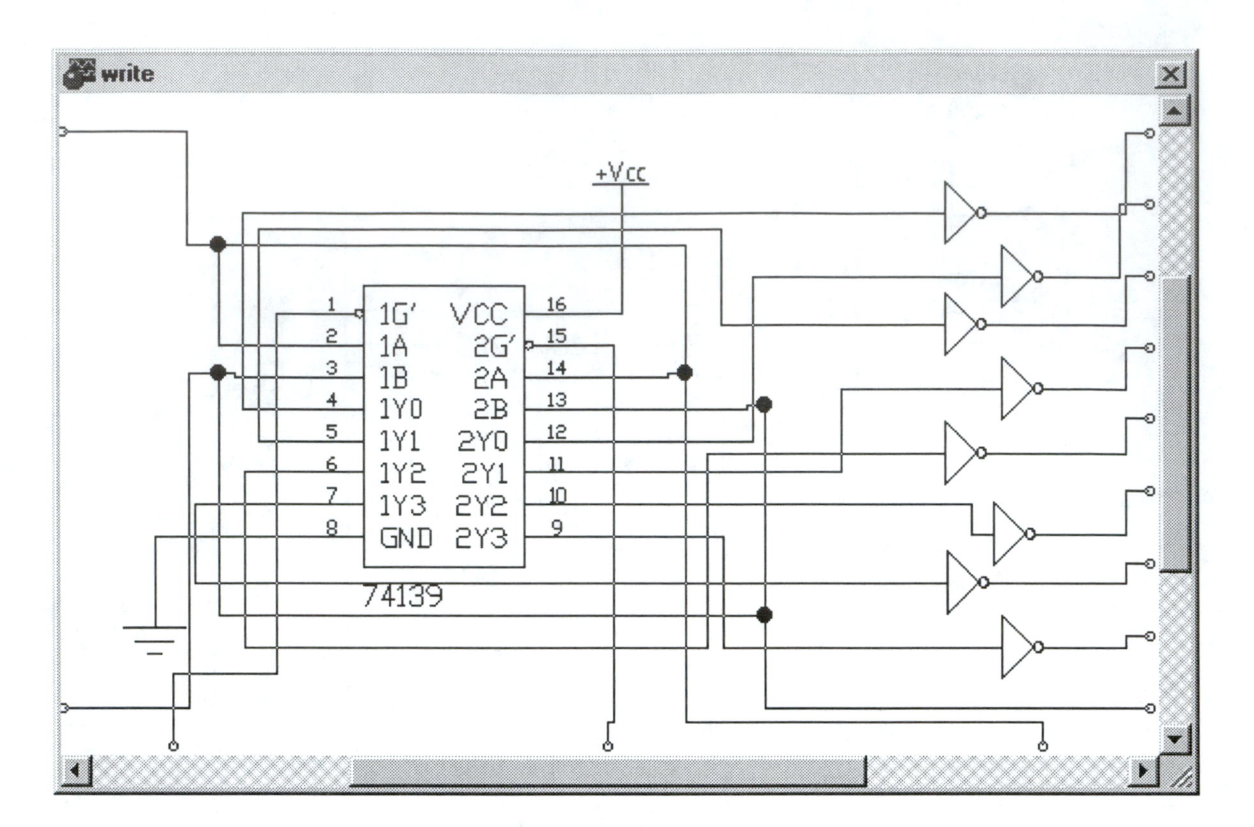

Figure 12.4: Write subcircuit

7. What is the 74139 used for? What purpose do the inverters serve?

8. Design a four-bit-wide memory. All SR latches get updated during a write, or examined during a read. Address lines should select one of four groups of bits.

9. Redesign the memory of step 8 by replacing the SR latches with the 7475 four-bit latch shown in Figure 12.5.

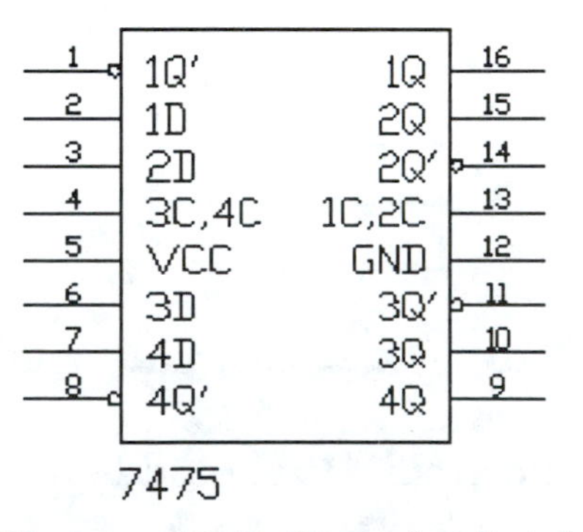

Figure 12.5: Four-bit latch

10. Use eight 7475's to make a four-byte memory. Read/write eight bits at a time.

11. Load the circuit **E12-2**, shown in Figure 12.6.

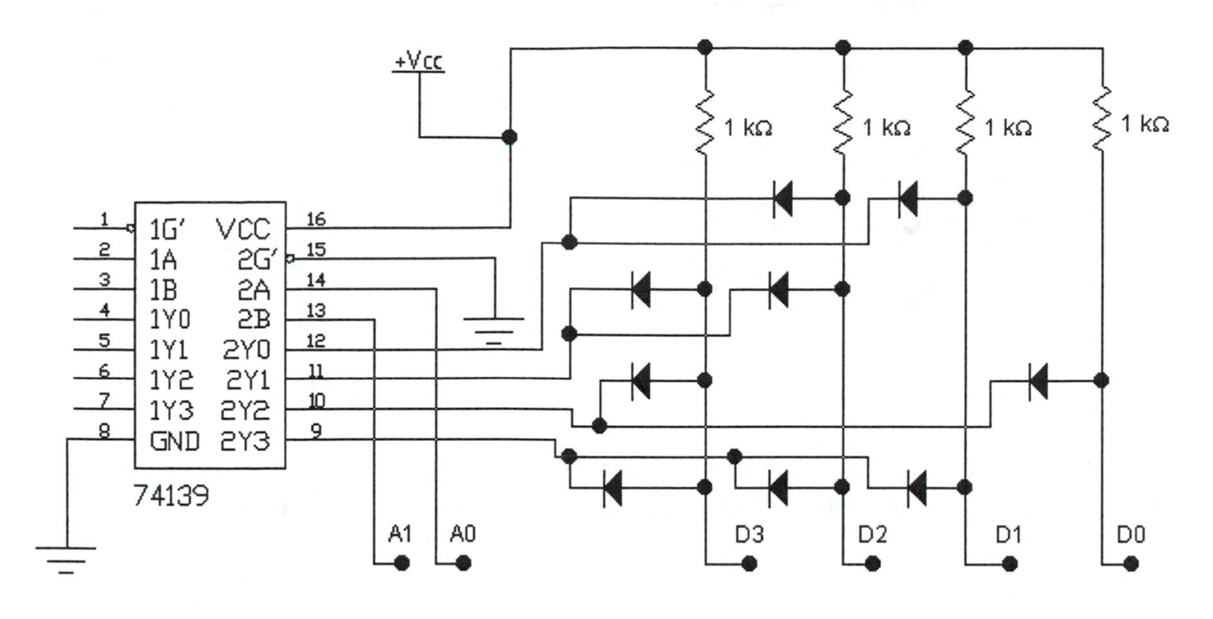

Figure 12.6: 16-bit diode-matrix ROM

Examine the operation of the diode-matrix ROM by connecting logic indicators to the data outputs and applying all address combinations to the address inputs. Fill in Table 12.3 during testing.

A1	A0	D3	D2	D1	D0
0	0				
0	1				
1	0				
1	1				

Table 12.3: ROM output

12. Use the other half of the 74139 to add four more locations to the ROM. Now three address lines are required to access a location. The additional four patterns should be 0100, 1110, 1111, and 1010.

13. *Troubleshooting:* Load the circuit **E12-3**, shown in Figure 12.7. This circuit is called an *address decoder*.

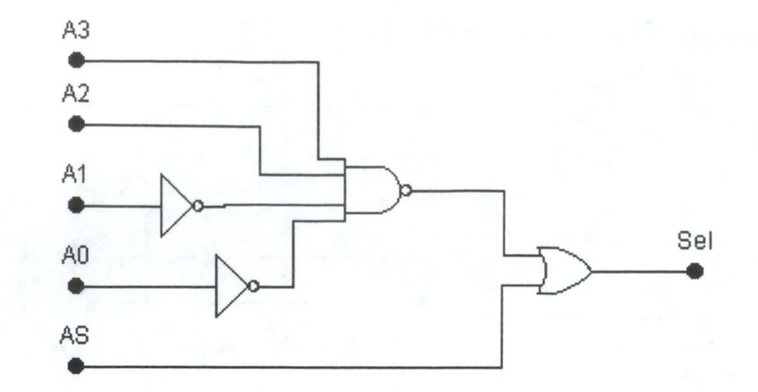

Figure 12.7: Simple address decoder

The Sel output should go low only when AS is low and the four-bit pattern 1100 is applied to the A0 – A3 inputs. Does the circuit work properly?

Discussion

While reviewing your data and results, provide detailed answers to each of the following:

1. How are 0s and 1s stored using latches?

2. What are decoders used for inside a RAM?

3. What are selectors used for inside a RAM?

4. How many 7475 4-bit latches are needed for a storage array containing eight rows of 32 bits? What type of decoders and/or selectors are needed to read/write any group of 16 bits?

5. What is the purpose of each diode in a diode-matrix ROM?

Experiment 13

Building Counters with Flip Flops

Name ______________________ **Date** __________

Objectives

The objectives of this experiment are to:

- Examine the operation of binary ripple counters.
- Learn how to use modulo-N counters.

Introduction

The toggle action of a flip flop can be used to implement a counting sequence. In this experiment, we will use D-type and JK flip flops to make ripple and modulo-N counters. A ripple counter is constructed by cascading multiple flip flops together in such a way that the Q output of the previous flip flop acts as the clock for the following flip flop. A modulo-N counter is a counter that has N states, a counting sequence that goes from zero to N–1.

Procedure

1. Load the circuit **E13-1**, shown in Figure 13.1.

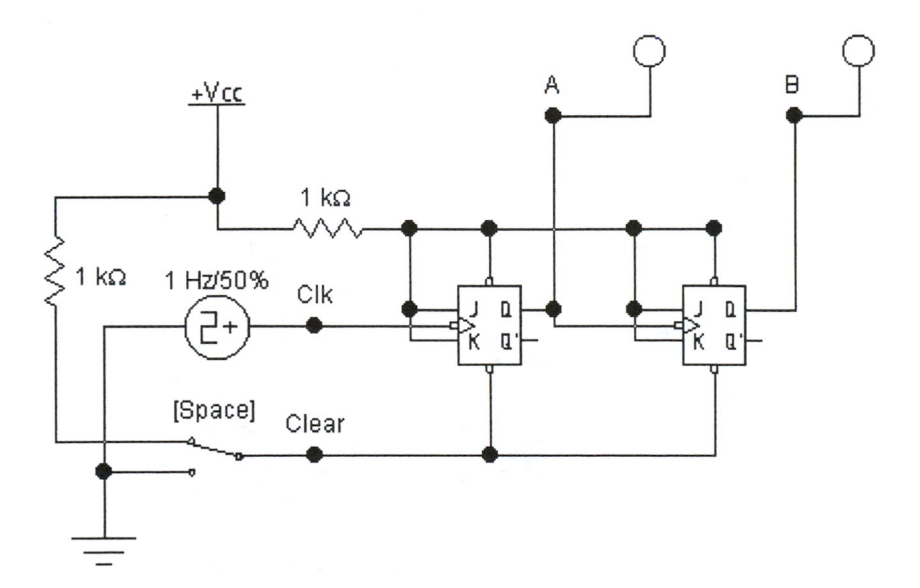

Figure 13.1: Two-bit binary ripple counter

2. The A output is the least significant bit. Simulate the circuit and note the activity of the logic indicators. Fill in Table 13.1 with the patterns.

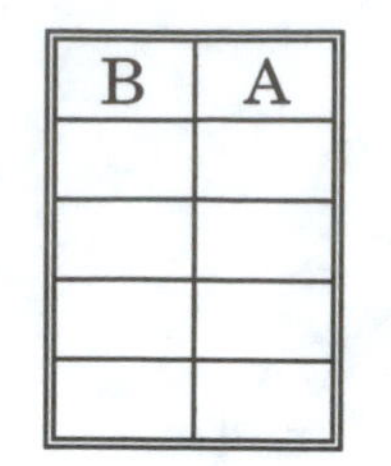

B	A

Table 13.1: Output patterns for 2-bit counter

3. What happens when the Clear input is forced low?

4. Connect the clock signal and both outputs to the logic analyzer and capture the waveforms for an entire counting sequence.

5. Add a third flip flop and assign the label C to its Q output . Simulate the circuit. How many patterns are in the counting sequence now? List them in order in Table 13.2.

C	B	A

Table 13.2: Output patterns for 3-bit counter

6. Add a fourth flip flop (output D), simulate the circuit, and record the output patterns in Table 13.3.

DCBA	DCBA	DCBA	DCBA

Table 13.3: Output patterns for 4-bit counter

7. Load the circuit **E13-2**, shown in Figure 13.2.

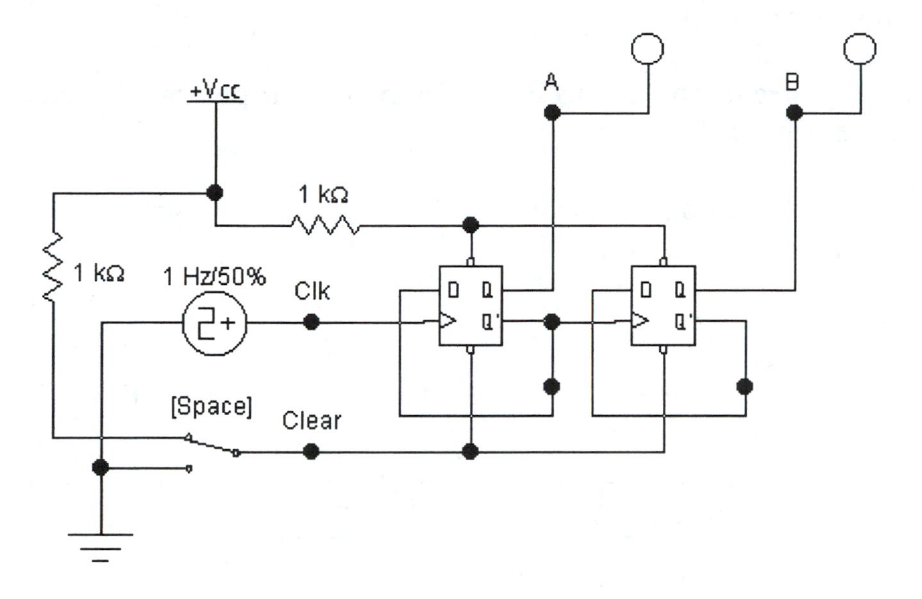

Figure 13.2: Ripple counter using D-type flip flops

Verify that the same 2-bit counting sequence is generated by the D-type flip flops.

8. Modify the circuit so that the Q output (not Q′) is used to clock the second flip flop. Does the counting sequence change?

9. Load the circuit **E13-3**, shown in Figure 13.3.

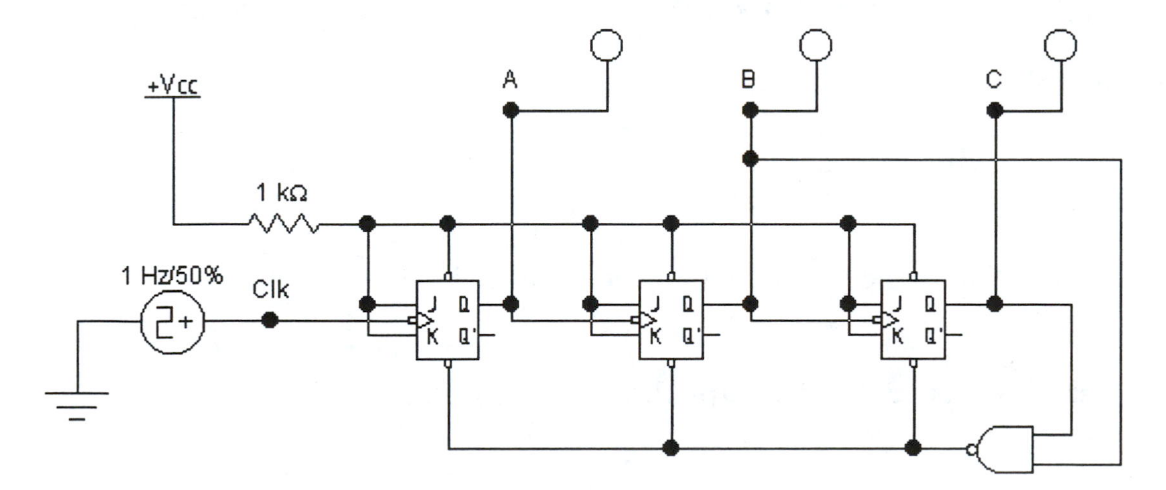

Figure 13.3: Modulo-6 counter

10. What does the counting sequence of the modulo-6 counter look like? Is it the same as the 3-bit binary counter?

11. Modify the circuit so that an active-high Clear input can be used to reset the modulo-6 counter.

12. *Troubleshooting:* Load the circuit **E13-4**, shown in Figure 13.4.

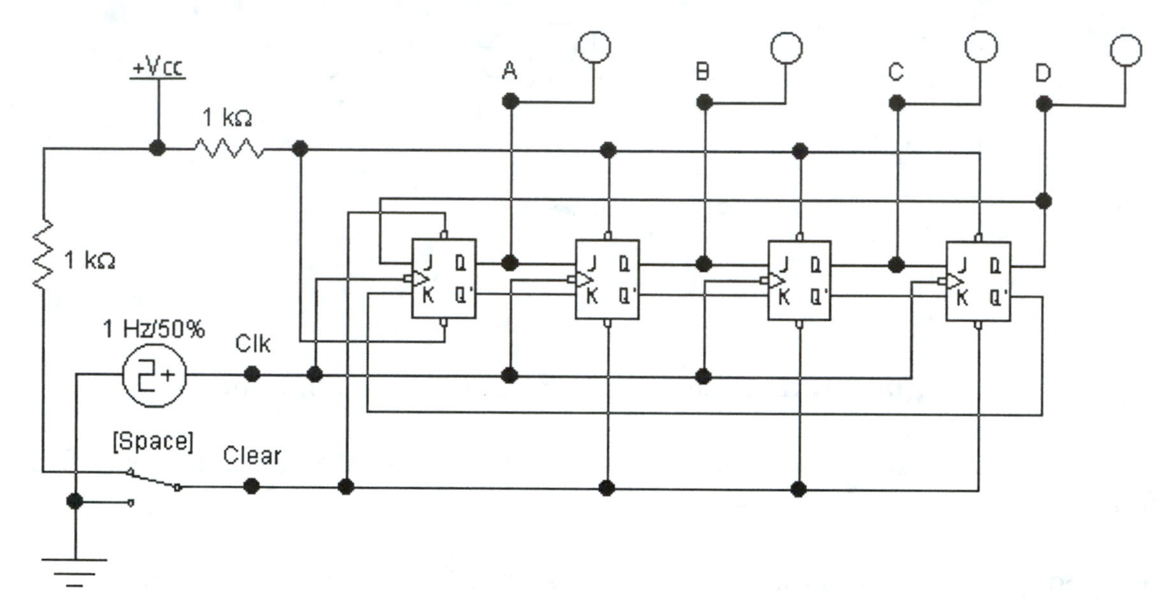

Figure 13.4: 4-bit ring counter

The 4-bit ring counter uses a shift register to generate the sequence 1000, 0100, 0010, 0001, repeatedly. Is the ring counter working correctly?

Discussion

While reviewing your data and results, provide detailed answers to each of the following:

1. How is toggle mode enabled on the JK and D-type flip flops?

2. Can the Q output of a D-type flip flop be used as the clock in a ripple counter?

3. How many patterns does a 5-bit ripple counter produce?

4. How can a modulo-5 counter be constructed?

Experiment 14

Binary and BCD Counters

Name ____________________ **Date** __________

Objectives

The objectives of this experiment are to:

- Examine binary and BCD counter packages.
- Cascade multiple counters for large counting sequences.

Introduction

In this experiment we examine several standard counting packages. The devices are used for both binary and BCD (binary coded decimal) counting. BCD counters are also called *decade* counters, since they only count from 0 to 9.

We will also see how multiple counters are cascaded to make larger counters, such as two 4-bit counters making an 8-bit counter.

Procedure

1. Load the circuit **E14-1**, shown in Figure 14.1.

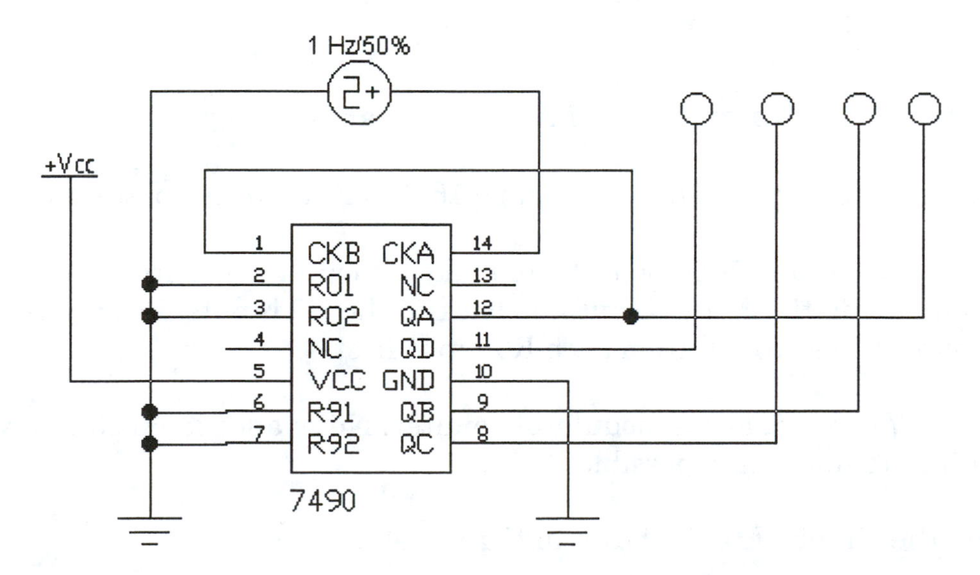

Figure 14.1: 7490 decade counter

The 7490 decade counter is wired internally to reset from nine back to zero. Verify that the counting sequence goes from zero to nine.

2. What happens if both R0 inputs are pulled high while the 7490 is counting? Demonstrate this in the circuit.

3. What happens if both R9 inputs are pulled high during counting?

4. What happens if the B-CLK input is disconnected from the A output during counting?

5. View the counting sequence using the logic analyzer.

6. Load the circuit **E14-2**, shown in Figure 14.2.

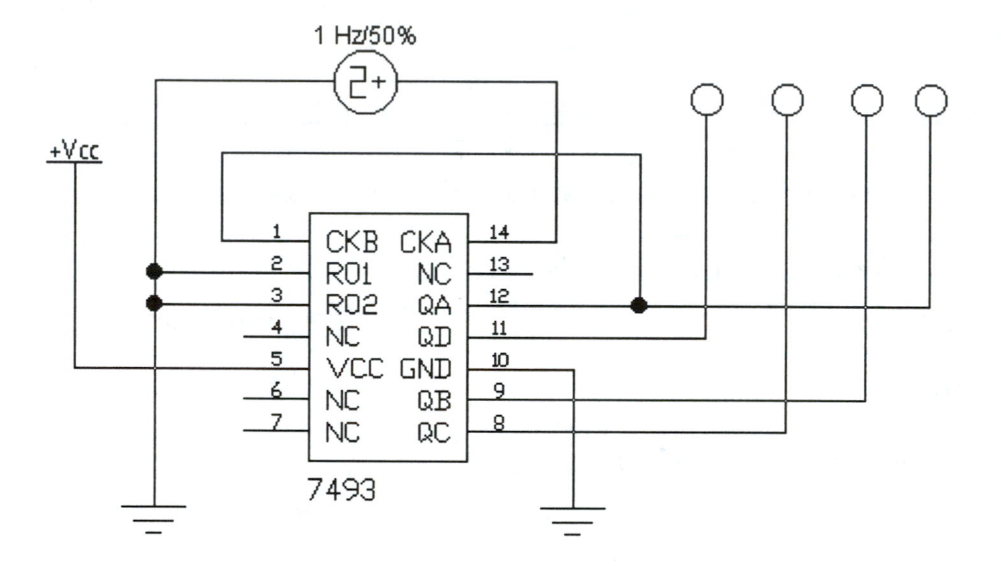

Figure 14.2: 7493 4-bit binary counter

7. Simulate the circuit. There should be 16 different output patterns.

8. Use either circuit (7490 or 7493) to make a modulo-6 counter. This is done by connecting the B and C outputs back to the R0 inputs. Be sure to disconnect the ground from each R0 input first.

9. Use the 7493 to make a modulo-11 counter. Note: additional circuitry is required to make this possible.

10. Load the circuit **E14-3**, shown in Figure 14.3.

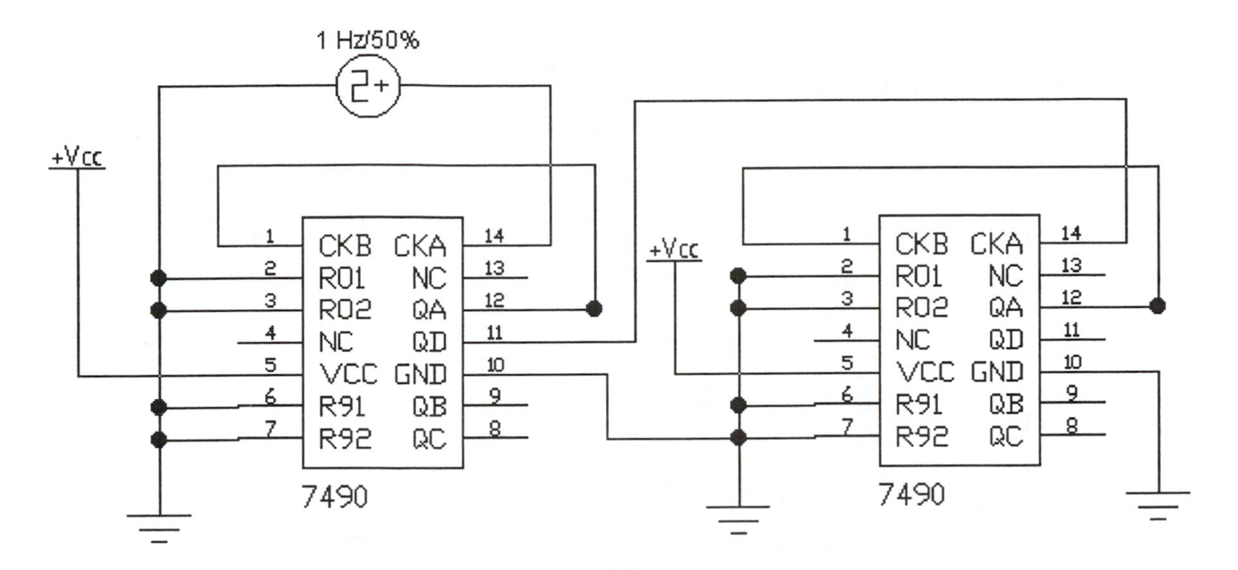

Figure 14.3: Cascaded decade counters

11. Use the Logic Analyzer to verify that there are 100 different output patterns. The A output on the first 7490 is the LSB. The D output on the second 7490 is the MSB.

12. If the input frequency is set to 1 kHz, what is the frequency on each of the eight outputs?

OUTPUT	FREQ	OUTPUT	FREQ
A1 (LSB)		A2	
B1		B2	
C1		C2	
D1		D2 (MSB)	

Table 14.1: Output frequencies for cascaded 7490's

13. Repeat steps 11 and 12 for two cascaded 7493 counters. How many output patterns are there? What are the frequencies at each output?

OUTPUT	FREQ	OUTPUT	FREQ
A1 (LSB)		A2	
B1		B2	
C1		C2	
D1		D2 (MSB)	

Table 14.2: Output frequencies for cascaded 7493's

14. Use two 7490's or two 7493's to make a modulo-87 counter.

15. *Troubleshooting:* Load the circuit **E14-4**, shown in Figure 14.4.

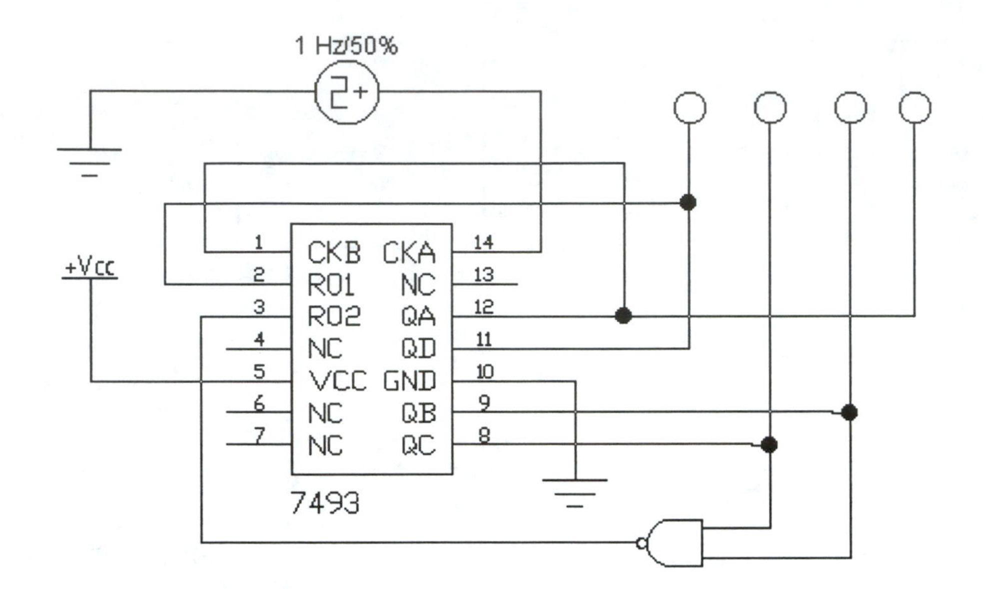

Figure 14.4: Modulo-14 counter

The counter should reset at 14. Does it?

Discussion

While reviewing your data and results, provide detailed answers to each of the following:

1. What clock edge is required by the 7490 and 7493?

2. How many 7490's are required to make a modulo-320 counter? How many 7493's?

3. Why is it proper to use the D output (the MSB) as the clock to the next counter when cascading counters?

4. Are the 7490 and 7493 counters pin compatible?

Experiment 15

The Seven-Segment Display

Name ______________________ **Date** __________

Objectives

The objectives of this experiment are to:

- Examine the operation of the seven-segment display.
- Use a BCD-to-seven-segment decoder to operate a seven-segment display.

Introduction

The seven-segment display, as shown in Figure 15.1, consists of seven light-emitting diodes arranged in a figure-eight pattern. Each LED illuminates a segment in the display. The cathode of each LED is wired to a common point, leaving seven anode connections to the display. This is called a common-cathode display. The cathode is typically connected to ground, and positive voltages are applied to the segment anodes for illumination. This is the type of display emulated by Electronics Workbench. Common-anode displays are also available.

By turning on different patterns of segments, we can form the digits zero through nine (and quite a few letters as well).

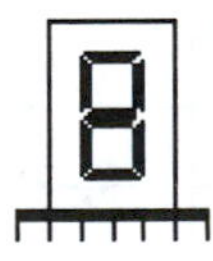

Figure 15.1: Seven-segment display

A prepackaged integrated circuit is available that converts a BCD pattern into the appropriate segment patterns to display the digits zero through nine.

Procedure

1. Load the circuit **E15-1**, shown in Figure 15.2.

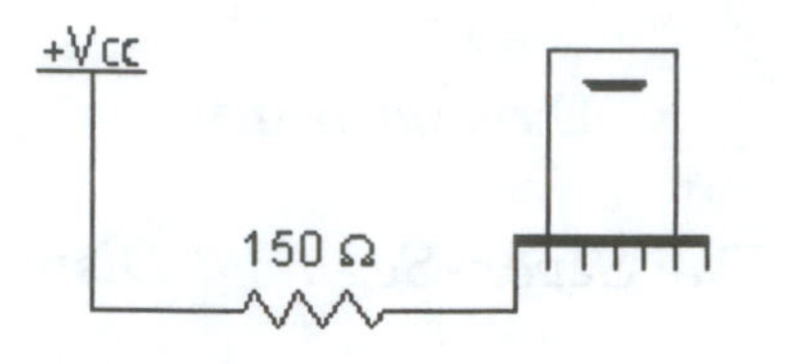

Figure 15.2: Segment A illuminated

The 150-ohm resistor is used to limit the segment current. The seven segments are labeled A through G, with segment A connected and illuminated.

2. Determine (by illumination) where segments B through G are located on the seven-segment display.

3. Connect the segments required to make the digit zero on the display.

4. Fill in Table 15.1 with the segment patterns associated with each digit.

DIGIT	A	B	C	D	E	F	G
0							
1							
2							
3							
4							
5							
6							
7							
8							
9							

Table 15.1: Segment patterns for each digit

5. Load the circuit **E15-2**, shown in Figure 15.3. A generic BCD-to-seven-segment decoder is used to drive the display. The decoder contains all logic required to turn the segments on in the appropriate patterns. Additional inputs are provided for special features.

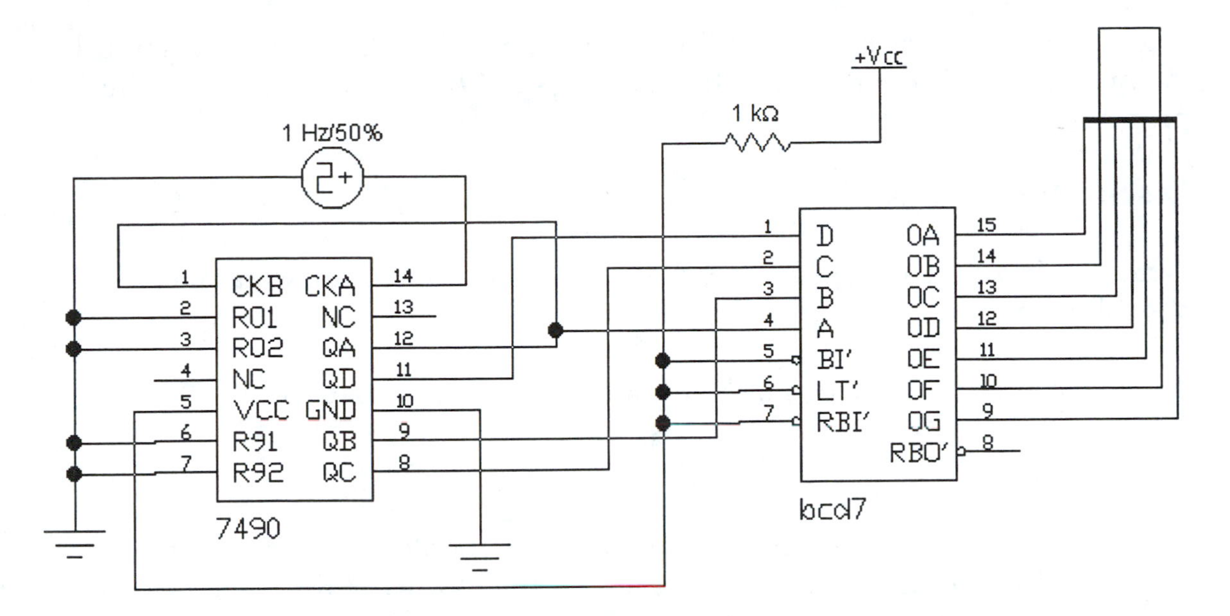

Figure 15.3: BCD-to-seven-segment decoder

6. Simulate the circuit to verify that the digits zero through nine show up on the display.

7. Determine what the LT′ input does and verify its operation.

8. Determine what the BI′, RBI′, and RBO′ signals are used for.

9. Replace the 7490 with a 7493 and resimulate the circuit.

10. Replace the generic BCD7 decoder in circuit E15-2 with a 7447 decoder from the 74xx Template parts bin. How does its operation compare to the generic decoder?

11. *Troubleshooting:* Load the circuit **E15-3**, shown in Figure 15.4.

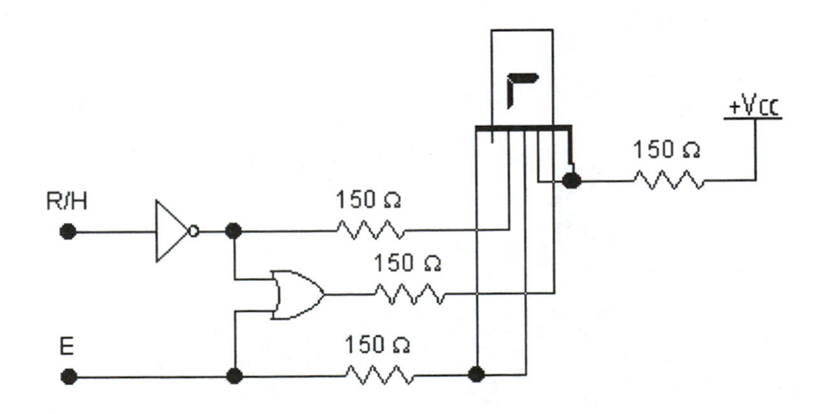

Figure 15.4: Status indicator

The status indicator should display an 'r' when R/H is high, an 'h' when R/H is low, and an 'E' when E is high, regardless of R/H. Is the circuit working properly?

Discussion

While reviewing your data and results, provide detailed answers to each of the following:

1. How could the 7447 be used with a common-anode display?

2. Set-up and solve a 4-input Karnaugh map capable of generating the bit pattern necessary to illuminate the A segment during a zero-to-nine count sequence. Use the number of logic gates in your solution to estimate the number of gates used inside the 7447.

3. Explain how to test a seven-segment display using the display decoder.

4. What letters (uppercase and lowercase) can be formed by the seven-segment display?

Experiment 16

The Digital Clock

Name ____________________ **Date** __________

Objectives

The objectives of this experiment are to:

- Examine how individual counting elements are combined to make a clock.

Introduction

In this experiment a complete digital clock is simulated by cascading modulo-10 and modulo-6 counters (with displays). The clock runs on military time (00:00:00 to 23:59:59) and allows hours and minutes to be set with the push of a button.

Procedure

1. Load the circuit **E16-1**, shown in Figure 16.1.

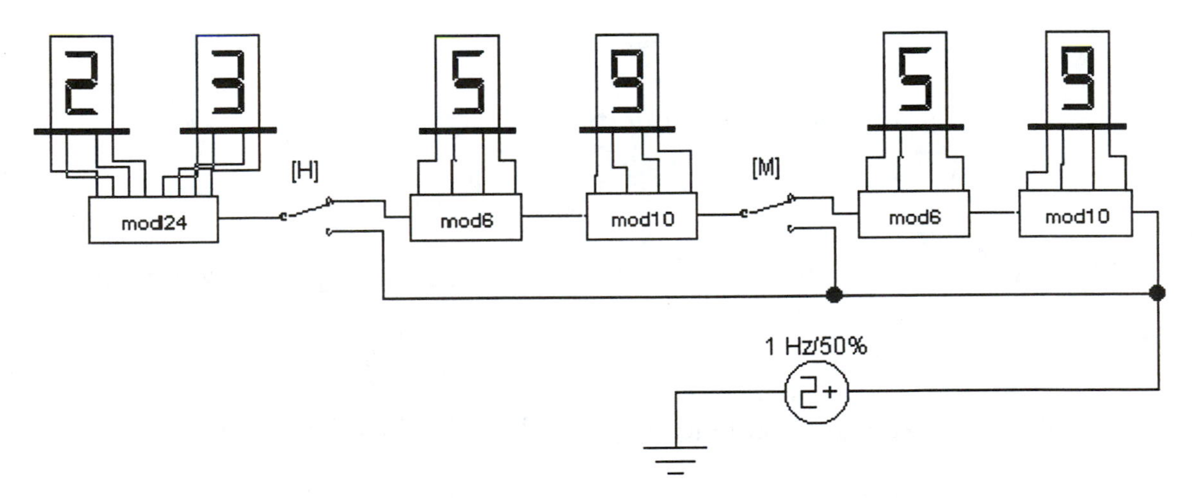

Figure 16.1: Digital clock

2. Simulate the circuit. The clock should reset to 00:00:00 and begin counting.

3. Use 'H' and 'M' to set the time to 23:58:00 (the seconds digits do not matter). Does the clock roll over properly (back to 00:00:00) after 23:59:59?

4. Examine the mod6 subcircuit. Is the AND gate really necessary to reset the 7490 to zero? Try eliminating the AND gate and feeding the B and C outputs back to the R0 inputs (use output C as the clock signal to the next stage). Does the clock work properly this way? Watch it for a full minute before deciding.

5. Examine the mod24 circuit. Once again, try eliminating the AND gate used to reset the counters. What happens?

6. *Troubleshooting:* Attempt to explain why the AND gate is needed in the mod6 subcircuits. Why does the direct connection from B and C back to the R0 inputs cause the counter to reset at '4'? Is there another way to reset the counter?

Discussion

While reviewing your data and results, provide detailed answers to each of the following:

1. How could leading zero blanking on the hours display (06 becomes 6, for example) be accomplished?

2. How could a simple alarm be added to the clock?

3. What is required to change the clock over to a 12 hour, AM/PM format?

4. How could a month/day calendar circuit be added to the clock?

Experiment 17

Shift Registers

Name ______________________ **Date** __________

Objectives

The objectives of this experiment are to:

- Examine how flip flops are combined to make shift registers.
- Examine the operation of a pseudorandom number generator.

Introduction

The same flip flops that can be used to make a counter can be used to make a shift register. The difference is how the flip flops are clocked and what mode they operate in. In this exercise we will examine several shift register circuits and applications.

Procedure

1. Load the circuit **E17-1**, shown in Figure 17.1.

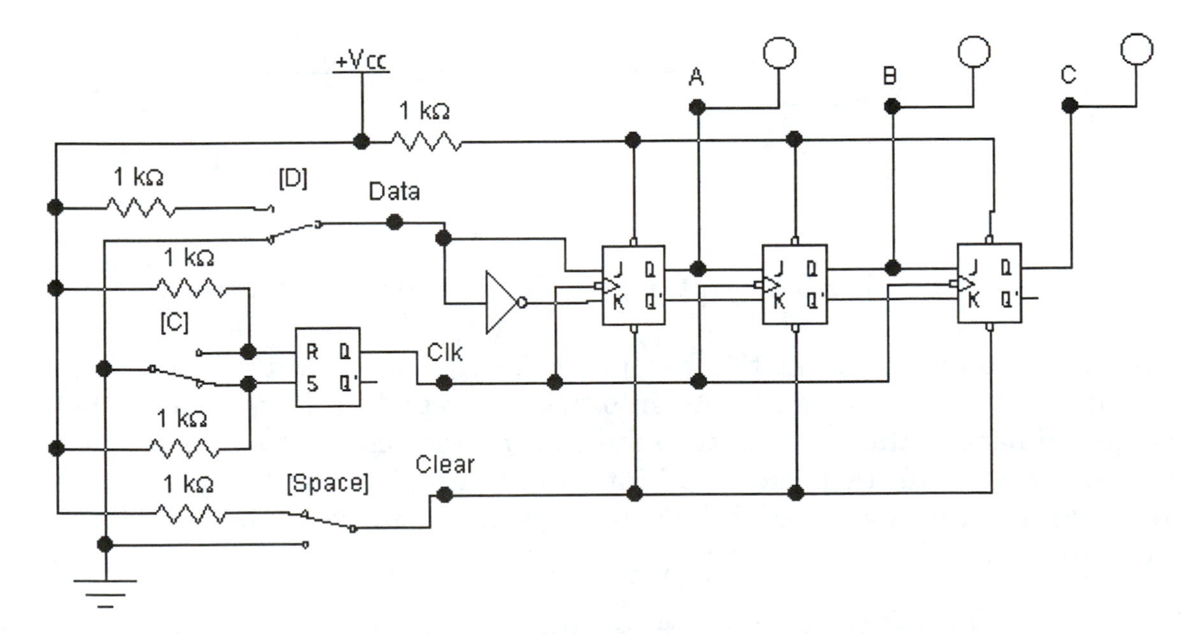

Figure 17.1: 3-bit shift register

2. Simulate the circuit. If the initial state is not 000, press the spacebar to clear the shift register. Note: press the spacebar twice to return it to its inactive state.

3. Use the 'D' switch to enter 0s and 1s into the shift register. To do this, set the Data switch to the desired position and then press 'C' twice to generate a clock edge. Practice loading a few bits into the shift register.

4. How many clock pulses does it take to fill the shift register with 1s after it has been cleared?

5. Load the circuit **E17-2**, shown in Figure 17.2.

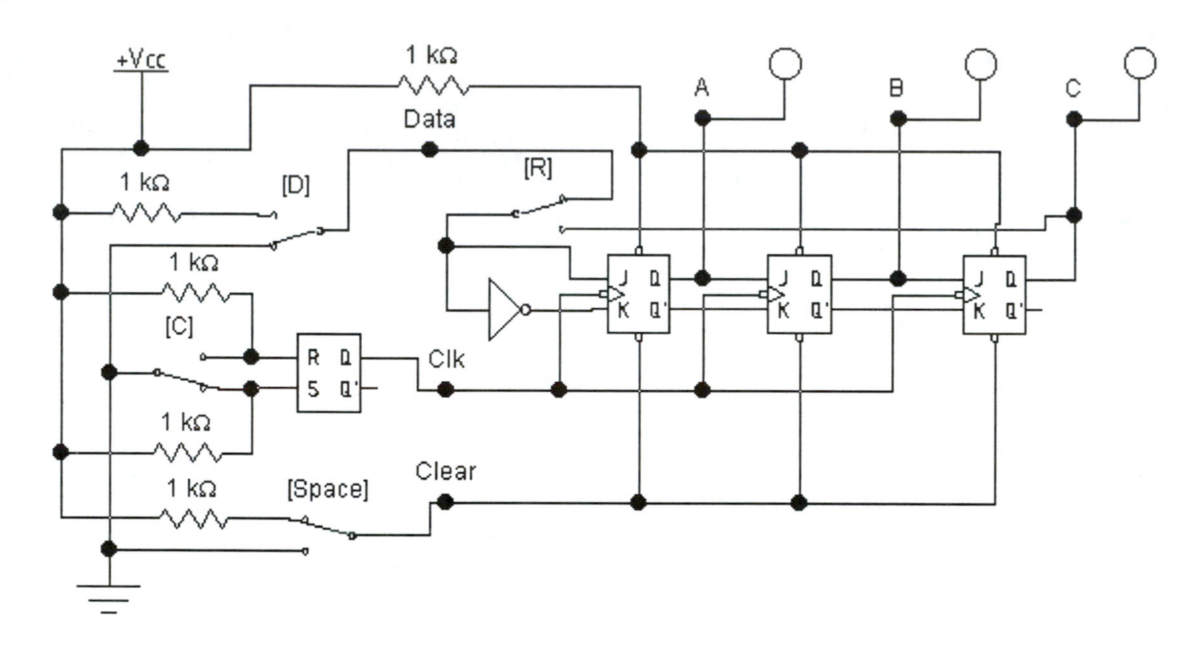

Figure 17.2: 3-bit shift register with recirculate control

This circuit is identical to E17-1 except for the addition of a *recirculate* switch. Once data has been loaded into the shift register, the output can be connected back to the input to keep the 3-bit pattern circulating inside the shift register. Load the pattern 110 into the shift register and verify the recirculate feature by clocking the shift register six times during recirculation.

6. Add a fourth flip flop to the shift register, then load and circulate the pattern 1011.

7. Load the circuit **E17-3**, shown in Figure 17.3.

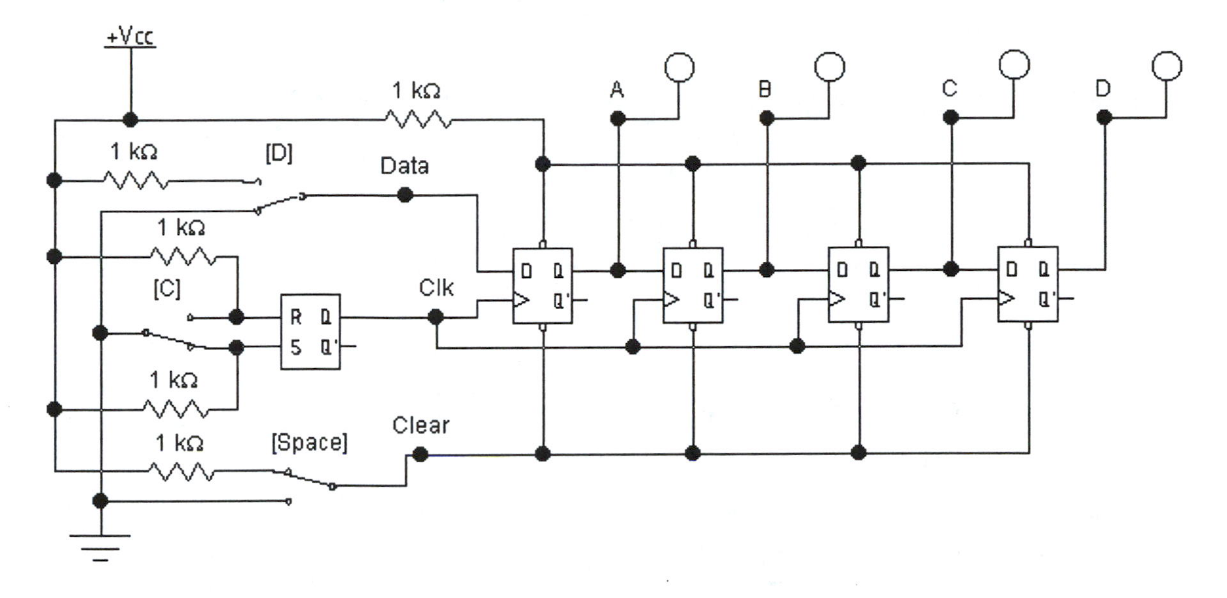

Figure 17.3: 4-bit D-type shift register

8. Compare the way data is clocked into the D-type shift register with that of the JK flip flop shift register.

9. Load the circuit **E17-4**, shown in Figure 17.4.

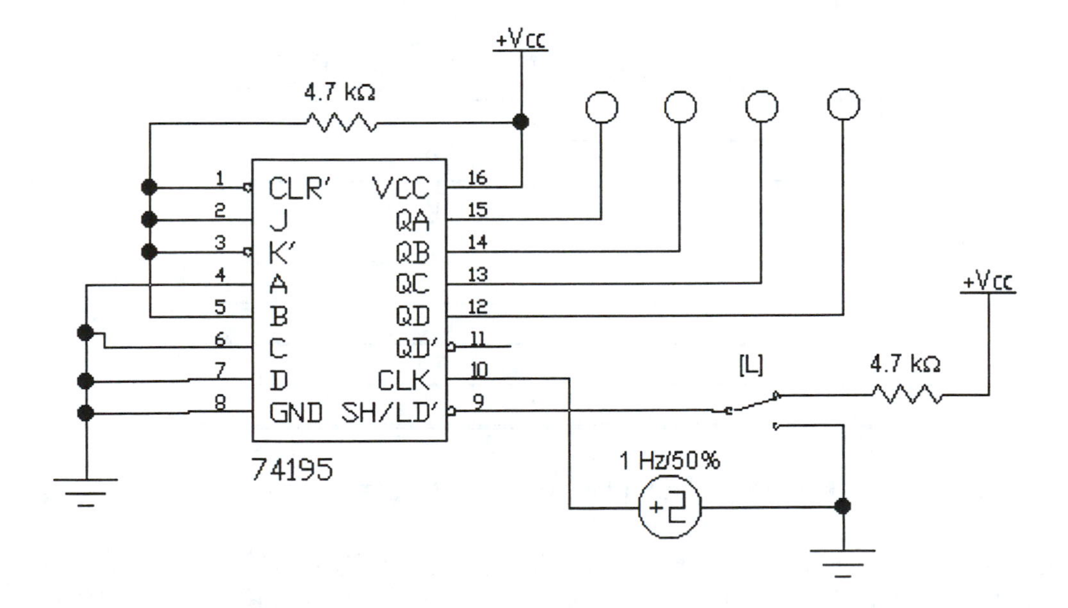

Figure 17.4: 74195 parallel-load shift register

10. Simulate the circuit. The shift register should eventually fill with 1s. When this happens, press 'L' once. Does the output instantly change to a new pattern? After a few seconds, press 'L' again. Now what happens?

Examine the help information for the 74195 to see how the shift/load input operates.

11. Load the circuit **E17-5**, shown in Figure 17.5.

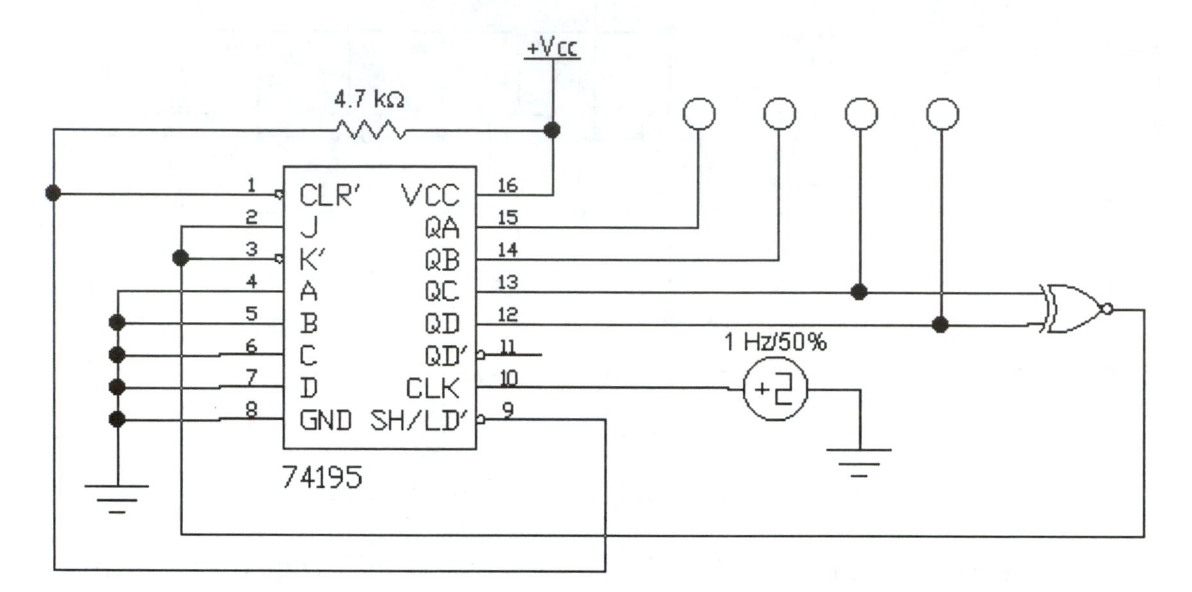

Figure 17.5: Pseudorandom pattern generator

12. Simulate the circuit. Record the output patterns in Table 17.1.

PATTERN	ABCD	PATTERN	ABCD
0		8	
1		9	
2		10	
3		11	
4		12	
5		13	
6		14	
7		15	

Table 17.1: Pattern generator output patterns

13. Load the circuit **E17-6**, shown in Figure 17.6. The 74165 allows an 8-bit number to be parallel loaded into the shift register (when the SH/LD′ input is low). Examine the circuit carefully, then try to predict what the output pattern will look like after using 'L' to load the shift register. Simulate the circuit and view the SH/LD′ input, the clock, and the serial output with the Logic Analyzer.

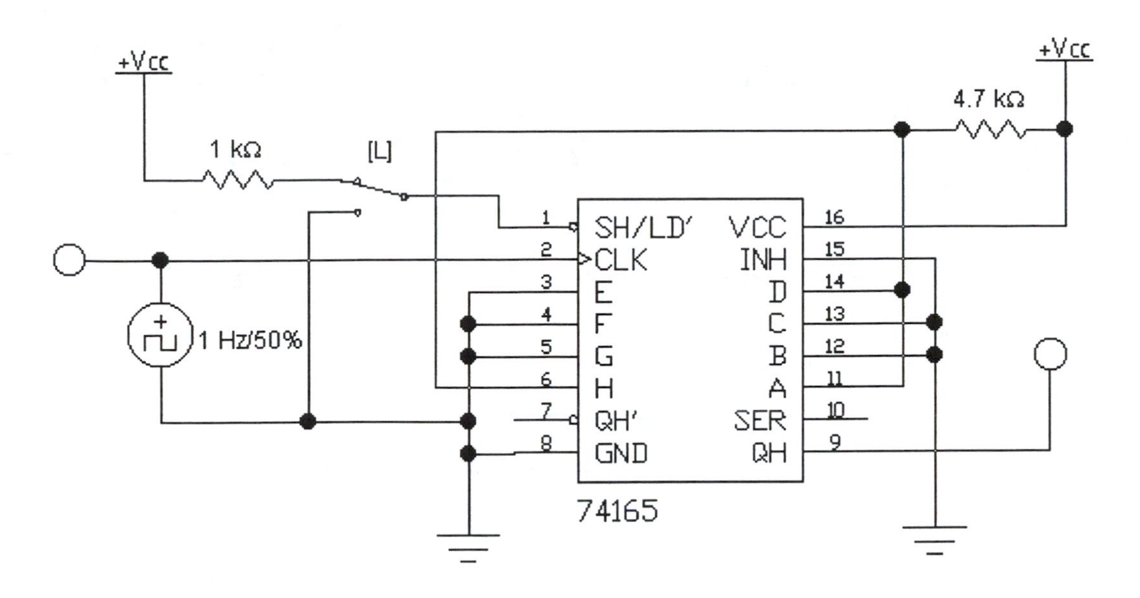

Figure 17.6: 74165 8-bit parallel-load shift register

14. *Troubleshooting:* Load the circuit **E17-7**, shown in Figure 17.7.

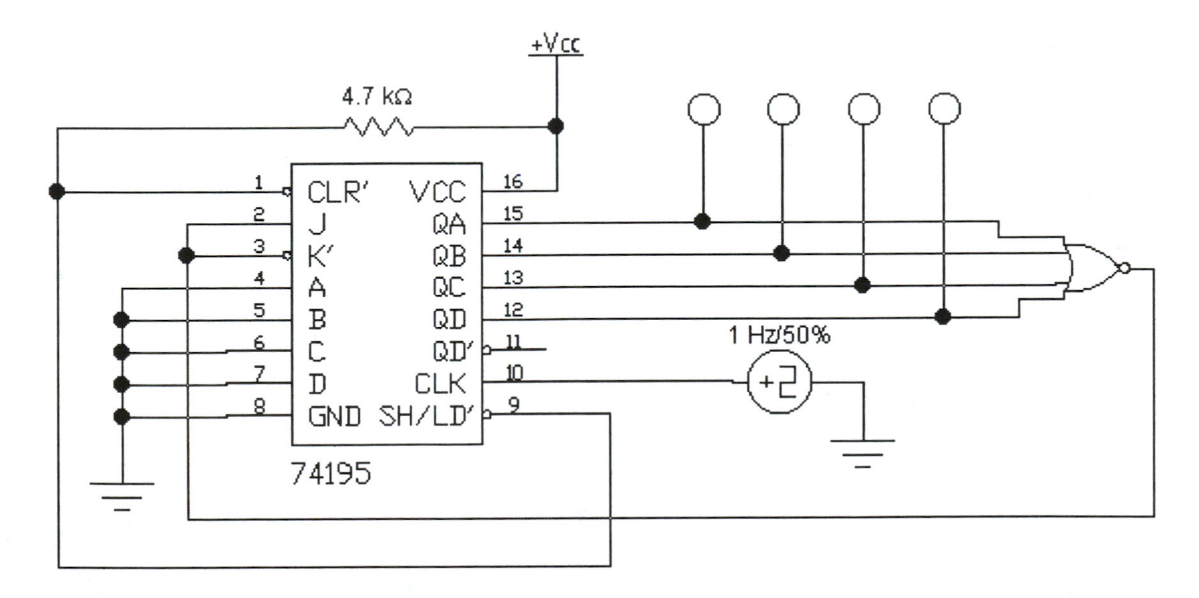

Figure 17.7: Troubleshooting circuit

The output should resemble that of a ring counter, with one output going high at a time. The expected pattern is 0000, 1000, 0100, 0010, 0001. Is the circuit working properly?

Discussion

While reviewing your data and results, provide detailed answers to each of the following:

1. Why does the pattern in a shift register shift only one bit position each time it is clocked?

2. What pattern is not output by the pseudorandom pattern generator?

3. Which shift registers covered in this experiment require a clock pulse to load data?

4. Explain how to build a 20-bit shift register.

Experiment 18

The Digital Multiplier

Name ______________________ **Date** __________

Objectives

The objectives of this experiment are to:

- Examine the method used to multiply binary numbers.
- Examine the operation of binary multiplier circuitry.

Introduction

When multiplying two binary numbers, one number is treated as a control word and the other number is shifted and added in various ways to form the result. Each bit of the control word is interpreted as follows:

0: shift left
1: accumulate and then shift left

The digital multiplier examined in this experiment uses the accumulate and shift method to multiply two 4-bit numbers.

Procedure

1. Load the circuit **E18-1**, shown in Figure 18.1.

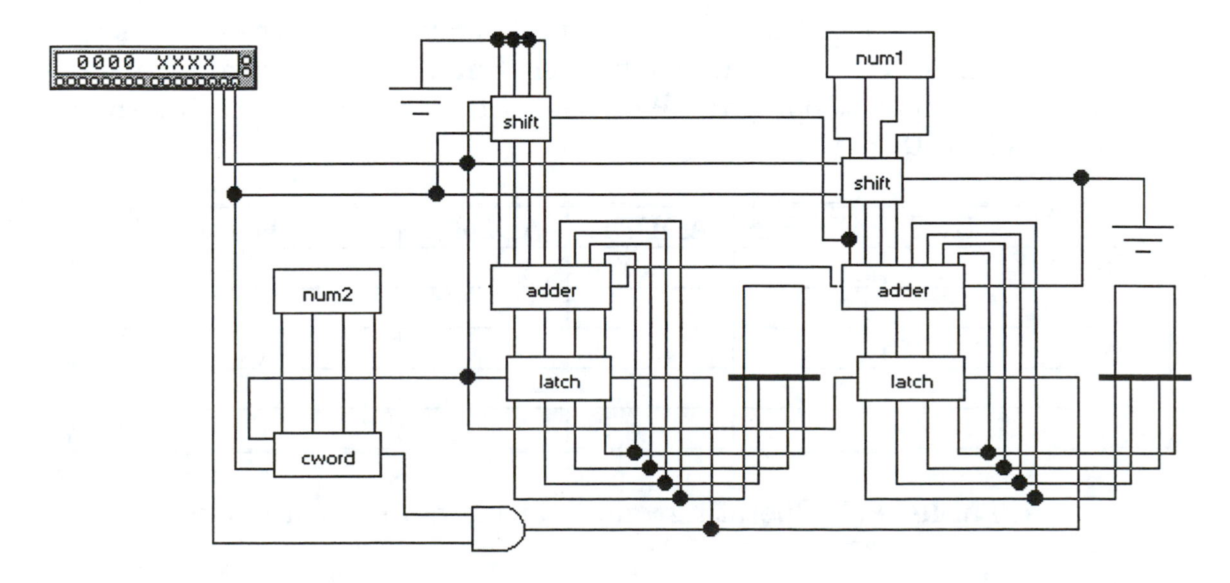

Figure 18.1: 4-bit by 4-bit binary multiplier

The 4-bit num1 value is multiplied by the 4-bit num2 value. The 8-bit result is displayed on the seven-segment displays. The num1 and num2 subcircuits each contain four switches so that the input numbers can be changed easily. The Word Generator contains a short sequence of patterns that control the shifting and accumulating performed by the shift, adder, and latch subcircuits.

2. The initial values of num1 and num2 are both 1111 (all switches are in the same position within the subcircuits). Simulate the circuit. The displays should start at 00, then become 0F, 2D, 69, and finally E1. Note that 1111 equals 15 decimal. Multiplying 15 by 15 gives 225, which equals E1 hexadecimal. Figure 18.2 illustrates this product.

```
    1111 (0F, first sum)
  +11110
  ------
  101101 (2D, second sum)
 +111100
 -------
 1101001 (69, third sum)
+1111000
--------
11100001 (E1, final result)
```

Figure 18.2: Accumulate and shift multiplication example

New sums are created whenever a one is shifted out of the cword subcircuit. Each sum is a combination of the current sum stored in the latch subcircuits and the pattern at the output of the two shift subcircuits.

3. Add several more decoded seven-segment displays so that the outputs of each shift and latch subcircuit can be seen. Simulate the circuit again (possibly by single-stepping the Word Generator) and record each display value in Table 18.1.

STEP	SHIFT	SHIFT	ADDER	ADDER	LATCH	LATCH
0						
1						
2						
3						
4						

Table 18.1: Operating values during multiplication

4. Double-click the num1 subcircuit. Its contents should look like Figure 18.3.

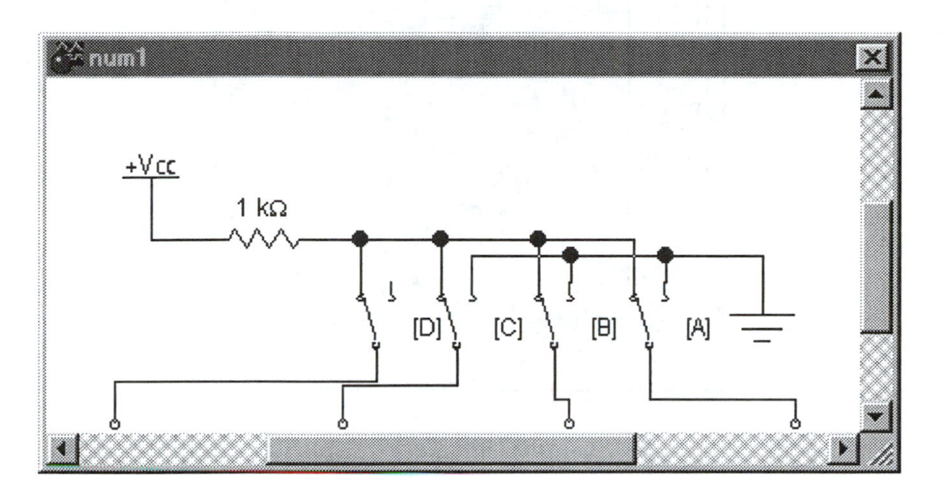

Figure 18.3: num1 subcircuit

5. Use these switches to change the 4-bit num1 pattern to 0101 and re-simulate. Is the product 4B?

6. Use the num2 subcircuit switches to set the control word to 0111 and re-simulate. Is the new product 23?

7. Find and record all products requested in Table 18.2.

*	0010	0011	0101	1100
0001				
0100				
0110				
1001				

Table 18.2: Test products

8. Multiply any num1 pattern by zero. Is the result zero as well?

9. Open the Word Generator. It should look like Figure 18.4.There are ten patterns used to control the sequence of events in the multiplier. The last pattern (0003) is set as the breakpoint so that the burst mode stops when the multiplier is finished. Stepping beyond the tenth pattern (to any 0000 pattern) clears the result, which is why the breakpoint is important.

   Pick any two input numbers from Table 18.2 and single-step through the ten patterns to find the product.

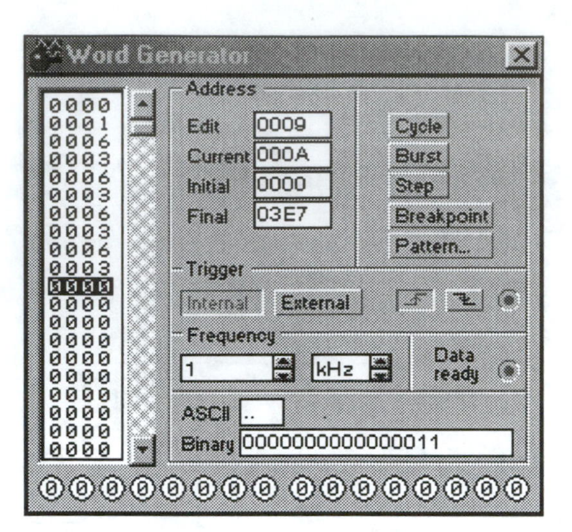

Figure 18.4: Word Generator

10. Connect the Logic Analyzer to the three Word Generator outputs and capture their waveforms during a multiplication (go back to burst mode and set the breakpoint again if necessary).

11. *Troubleshooting:* Design a combinational or synchronous logic circuit to generate the same sequencing waveforms as the Word Generator. Verify that the multiplier operates correctly with your sequencer.

Discussion

While reviewing your data and results, provide detailed answers to each of the following:

1. Verify that the products in steps 5 through 7 are correct.

2. What is the AND gate used for?

3. Why is the first shift subcircuit always loaded with 0000?

4. How can the multiplier be expanded to perform 8-bit by 8-bit multiplication?

Experiment 19

A / D Converters

Name ______________________ **Date** __________

Objectives

The objective of this experiment is to:

- Use an A/D converter to digitize an analog waveform.

Introduction

An Analog-to-Digital Converter (ADC) converts an analog voltage into a corresponding binary value. An 8-bit ADC converts a range of input voltages into one of 256 output patterns (00 to FF). Typically, an ADC has a digital input that triggers the start of a *sample*. When the ADC is finished converting the sample into binary, an end-of-conversion output signal changes state.

The input voltage range is specified by reference voltages applied to the ADC. For example, reference voltages of 0 V and 5 V indicate a 5-V range of input voltages, all above zero. Reference voltages of –2.5 V and +2.5 V set a 5V range as well, but allow dual-polarity signals to be sampled.

Electronics Workbench contains one analog-to-digital converter package, shown in Figure 19.1.

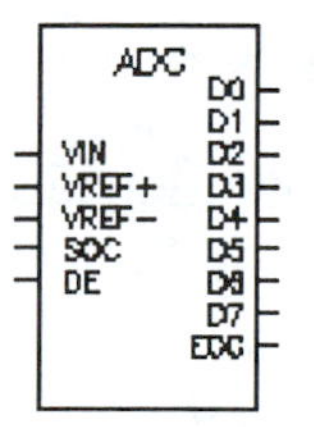

Figure 19.1: 8-bit ADC

A rising edge on SOC (start of conversion) triggers a conversion. The 8-bit output is available when EOC (end of conversion) goes high. If DE (data enable) is high, the outputs will indicate the converted sample value. When DE is low, the outputs are held at zero.

Procedure

1. Load the circuit **E19-1**, shown in Figure 19.2.

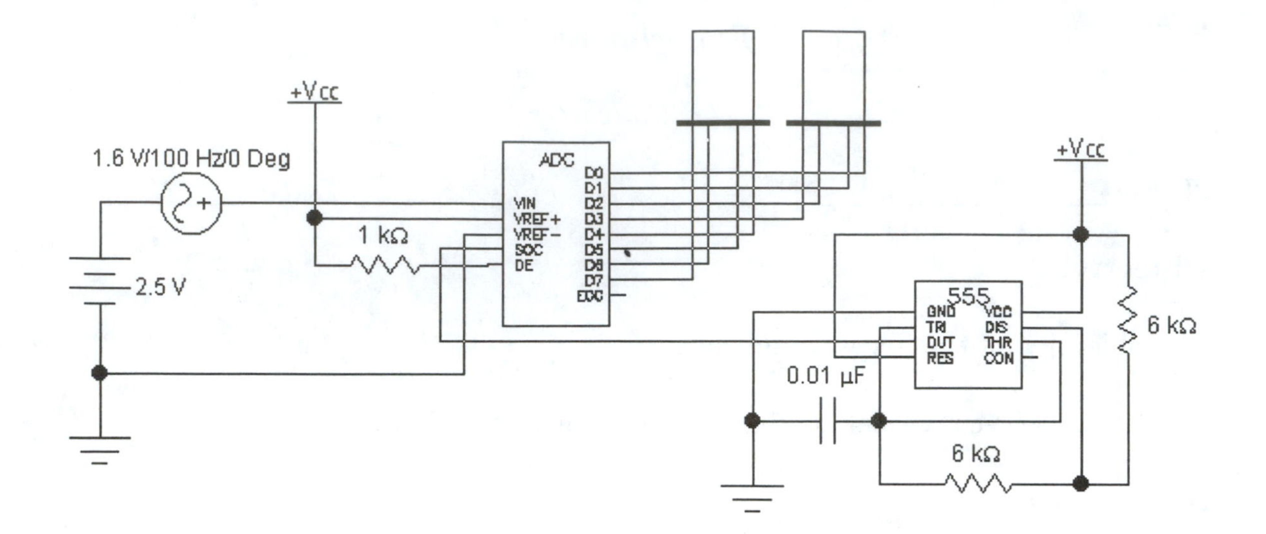

Figure 19.2: ADC test circuit

2. The input voltage at VIN is a 1.6 V rms, 100-Hz sinewave with a DC offset of 2.5 V. The positive and negative peaks of the input voltage are 4.76 V and 0.237 V (the negative peak is pushed above 0 V by the DC offset). These voltages fall within the 5-V range specified on the VREF+ and VREF− inputs. The 555 timer serves as the sample clock, causing the ADC to convert several thousand samples/second. Simulate the circuit. What happens on the displays? Record the smallest and largest hexadecimal values indicated.

SMALLEST	LARGEST

3. View the input voltage on the oscilloscope. Note the relationship between the input voltage and the displayed conversion value.

4. Increase the rms input voltage so that the waveform has a peak-to-peak voltage closer to 5 V. Record the new output range.

SMALLEST	LARGEST

5. Increase the DC offset voltage to 3 V. What happens at the output when the input voltage goes above VREF+?

6. Decrease the DC offset voltage to 2 V. What happens when the input goes below 0 V?

7. What is the sample rate of the ADC?

SAMPLE RATE

8. How many samples per cycle are taken?

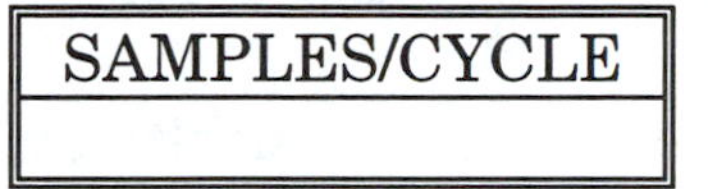

9. Load the circuit **E19-2**, shown in Figure 19.3.

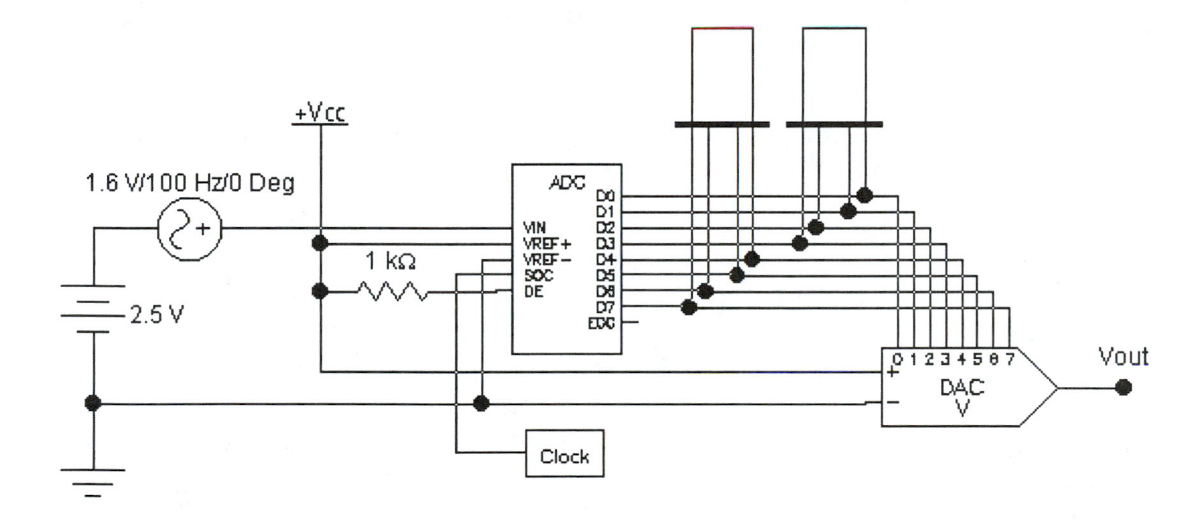

Figure 19.3: Converting ADC output back into analog

10. Use the oscilloscope to view the input and output waveforms at the same time. The DAC is a digital-to-analog converter that converts its binary input into a corresponding voltage within the range of voltages specified by the + and – inputs.

11. Open the Clock subcircuit and change both 6-k resistors to 60 k. Repeat step 10.

12. *Troubleshooting:* A circuit similar to E19-2 has an output waveform that looks like Figure 19.4. What is causing the distortion? Experiment with E19-2 until you can reproduce the distortion.

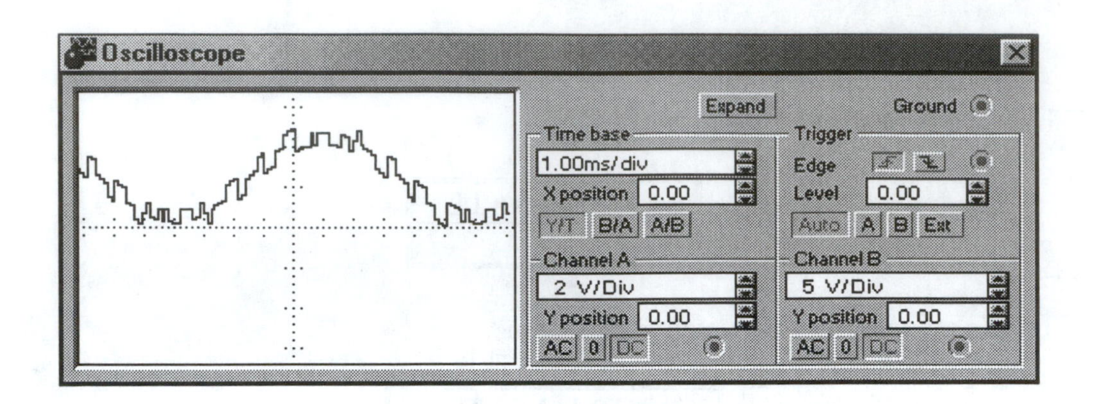

Figure 19.4: Troubleshooting waveform

Discussion

While reviewing your data and results, provide detailed answers to each of the following:

1. How does the sample rate affect the quality of the digitized waveform?

2. Which output value represents a more-positive sample, 3F or 4E?

3. What would happen in each circuit if the DC offset on the input voltage was eliminated?

4. What sample rate is required to obtain 25 samples/cycle of an 80-Hz waveform?

Experiment 20

D / A Converters

Name ____________________ **Date** __________

Objectives

The objectives of this experiment are to:

- Examine the operation of a digital-to-analog converter.
- Create simple waveforms using D/A converters.

Introduction

The digital-to-analog converter is a circuit that converts a binary input value (4 bits, 8 bits, or more) into a proportional voltage (or current). This process is illustrated in Figure 20.1.

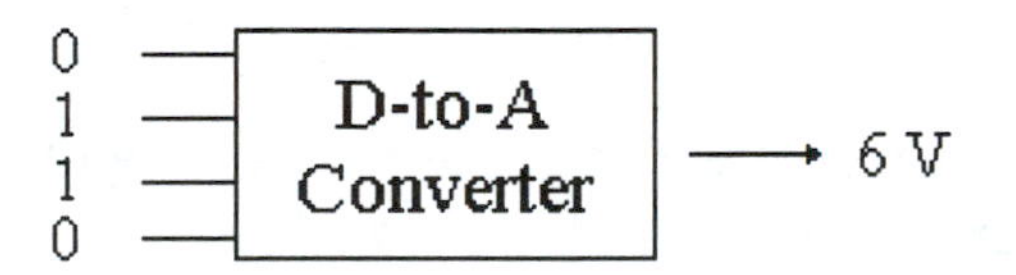

Figure 20.1: D/A converter

The least significant bit of the input controls the size of the *step voltage*. This is the minimum amount of voltage the output can change by. The range of output voltage for the D/A converter equals the product of the step size and the number of steps (the number of different input patterns). The range can be found by the following formula:

$$Vrange = Vstep \cdot (2^N - 1)$$

This formula can be used in reverse to find the required step size given a known voltage range.

Procedure

1. Load the circuit **E20-1**, shown in Figure 20.2. The op-amp is wired as a summer, with the gain of each input set to the appropriate binary weights. Since the summer configuration is inverting by nature, the output voltage will be negative, since the inputs tied to Vcc will be sitting at positive 5 volts.

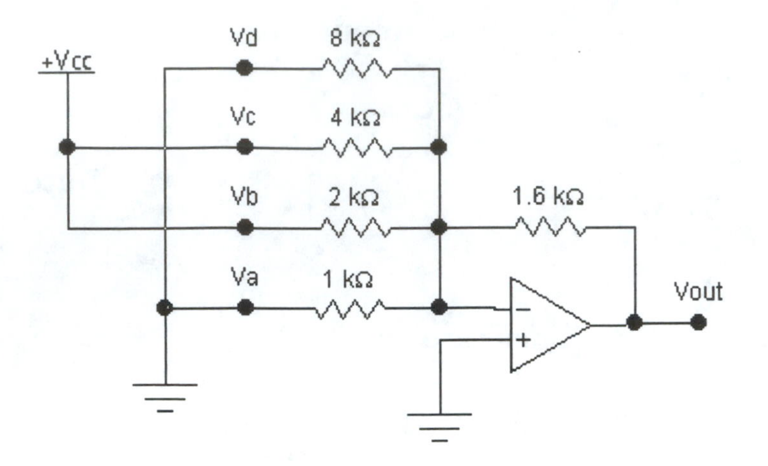

Figure 20.2: 4-bit D/A converter

2. Connect a DC voltmeter to the output and record Vout.

Vout

3. One by one, change each voltage source to its opposite voltage (0 V or 5 V) and record Vout each time.

Va	Vb	Vc	Vd	Vout

4. Load the circuit **E20-2**, shown in Figure 20.3. A decade counter is used to provide the input patterns, which results in a specific type of waveform at the output of the D/A converter.

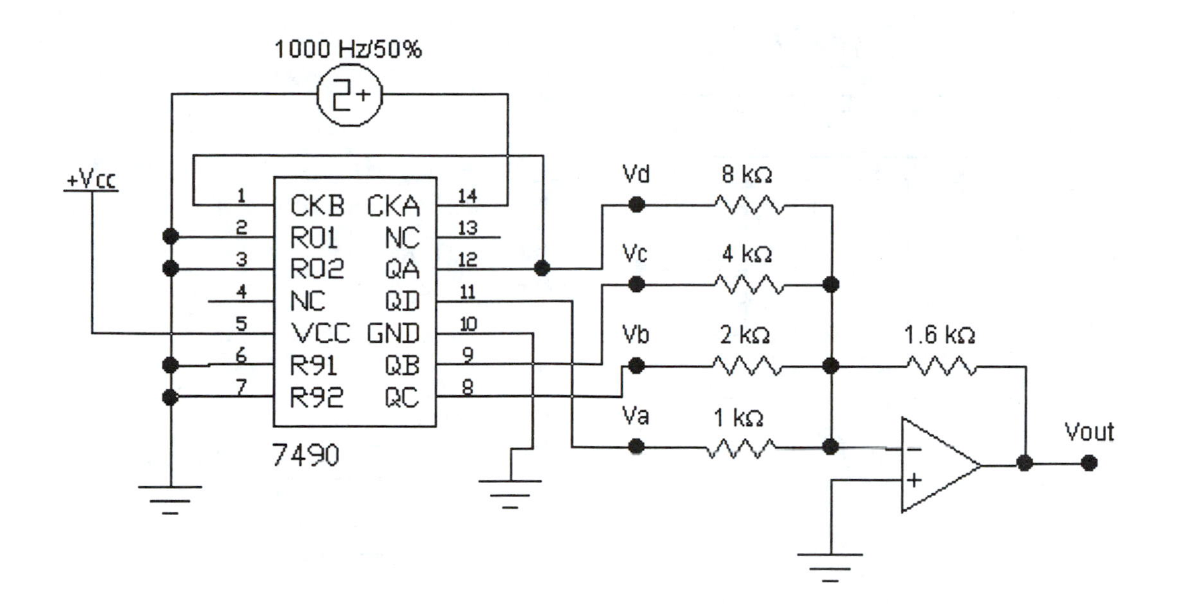

Figure 20.3: 4-bit D/A converter with decade counter

5. Connect the oscilloscope to display the main clock (CKA) and Vout .

6. What does the resulting waveform look like?

7. What is the range of output voltages?

Vout (min)	Vout (max)

8. Change the 7490 to a 7493 binary counter and repeat steps 6 and 7.

Vout (min)	Vout (max)

9. *Troubleshooting:* Load the circuit **E20-3**, shown in Figure 20.4. The range of output voltages should be 0 V to – 4.5 V. Does the troubleshooting circuit satisfy this range? If not, why not? Is the shape of the waveform correct? If not, determine why.

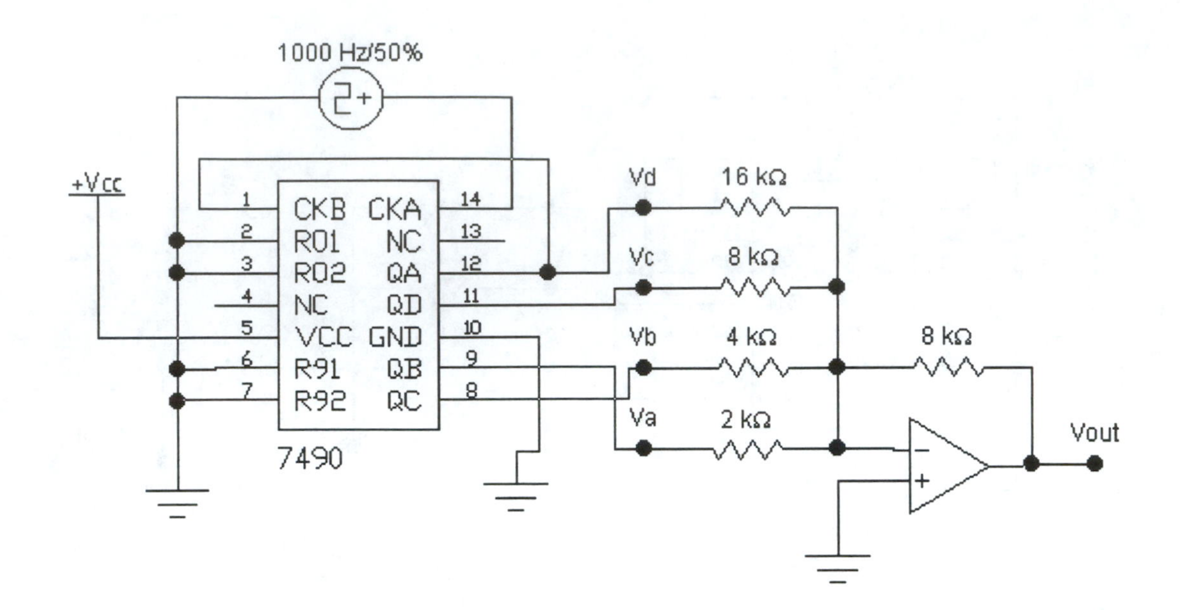

Figure 20.4: Troubleshooting circuit

Discussion

While reviewing your data and results, provide detailed answers to each of the following:

1. In the first two circuits, why is the feedback resistor 1.6 K ohms?

2. What changes must be made to the 4-bit D/A to make a 5-bit D/A?

3. Is there a limit as to how fast the D/A converter can be clocked? If so, what causes it?

4. How can positive voltages be generated by the 4-bit D/A?

Experiment 21

The Digital Voltmeter

Name ____________________ **Date** __________

Objectives

The objective of this experiment is to:

- Examine how a simple digital voltmeter is constructed using a counter, D/A converter, and comparator.

Introduction

The digital voltmeter examined in this experiment measures an unknown input voltage between zero and nine volts. A single seven-segment display indicates the measured voltage. A counter drives the display and also a 4-bit digital-to-analog converter (DAC). The output of the DAC goes from 0V to –9V in –1V steps and is compared with the unknown input voltage at each step. When the DAC output is greater than the input voltage the counter is stopped and the display indicates the measured voltage.

Procedure

1. Load the circuit **E21-1**, shown in Figure 21.1.

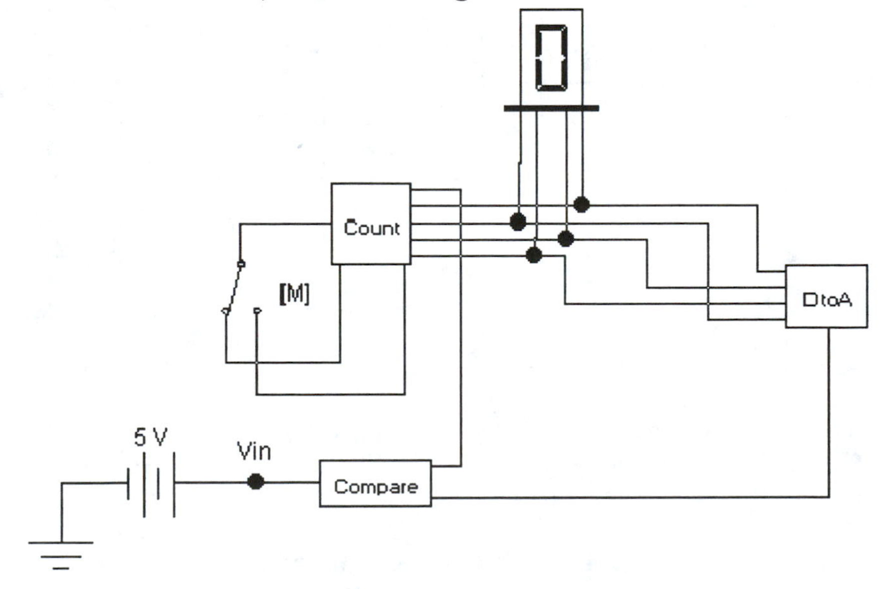

Figure 21.1: Digital voltmeter

2. The value of the unknown input voltage at Vin is set by the DC source. At the beginning of simulation, the display should read zero. Press 'M' to begin a measurement. The digital voltmeter will count up until the unknown voltage is found and then stop. Pressing 'M' a second time resets the digital voltmeter.

   Make all the measurements indicated in Table 21.1.

VIN (V)	DISPLAY	VIN (V)	DISPLAY	VIN(V)	DISPLAY
0		3		6	
1		4		7	
2		4.9		8	
2.1		5		8.5	
2.9		5.1		9.1	

Table 21.1: Sample measurements

3. Set Vin to 20 volts and measure it. What happens?

4. Open the Count subcircuit. It should look like Figure 21.2.

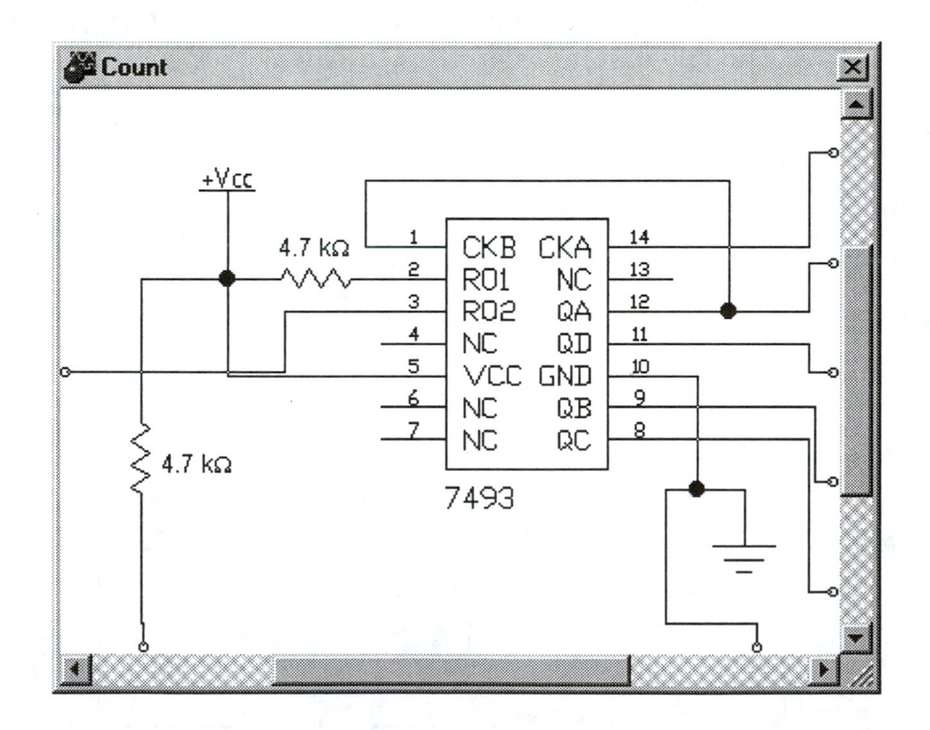

Figure 21.2: Count subcircuit

5. How does 'M' control the counting circuit?

6. Open the DtoA subcircuit. It should look like Figure 21.3.

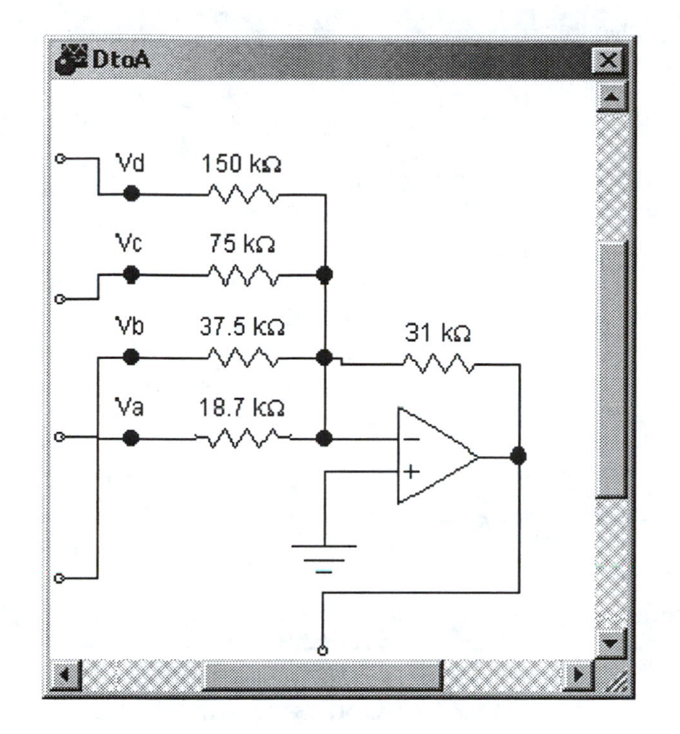

Figure 21.3: DtoA subcircuit

7. One input at a time, determine the gain and the contribution towards the output voltage (Vo). The first input (Vd) is analyzed as follows:

$$Vo = 5v\left(-\frac{31k}{150k}\right) = -1.03v$$

Where the logic one input voltage at Vd is 5 volts.
Fill in Table 21.2 with your calculations.

Va	Vb	Vc	Vd
−1.03v			

Table 21.2: DAC voltages

8. Open the Compare subcircuit. It should look like Figure 21.4. The 10 k ohm resistors are used as a voltage divider. The op-amp compares the divider output against zero. Normally the op-amp's output is high, which allows the clock signal to pass through the AND gate and increment the counter. When the DAC output 'equals' Vin, the output of the op-amp switches low, stopping the clock.

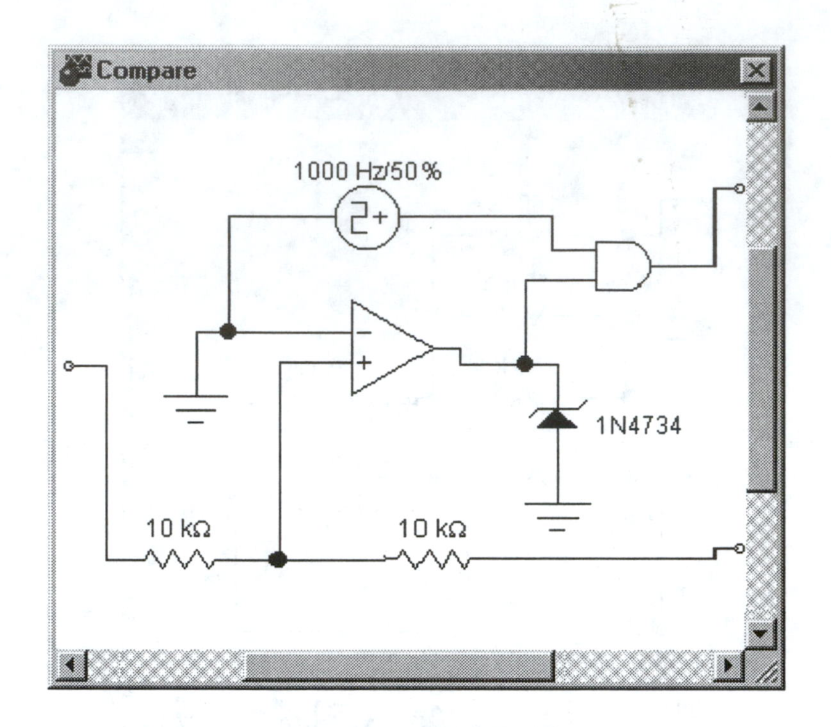

Figure 21.4: Compare subcircuit

9. Change the resolution of the DAC to –0.5 volts/step. Add a second display that only displays a zero or a five, depending on the output of the counter.

10. *Troubleshooting:* Modify the original circuit so that it stops automatically after one complete cycle of counting (the counter goes from 1111 to 0000).

Discussion

While reviewing your data and results, provide detailed answers to each of the following:

1. What is the Zener diode used for in the Compare subcircuit?

2. Explain how the voltage-divider output goes to zero when the DAC output 'equals' Vin. Is the DAC voltage actually equal to Vin?

3. Why use a 7493 counter instead of a 7490 counter?

4. What happens when the input voltage is very large, as in 20 V?

Experiment 22

Synchronous Logic Circuits

Name ______________________ **Date** __________

Objectives

The objectives of this experiment are to:

- Examine the operation of synchronous logic circuits.
- Interpret and use state diagrams.

Introduction

A synchronous logic circuit is essentially a combinational logic circuit with a memory section. The current state of the circuit, together with additional input signals, determines the next state. The state of the synchronous logic circuit (also called a *state machine*) changes only when a clock pulse is applied. Figure 22.1 shows a sample *state diagram*, which contains all of the states used by a machine and the transitions between the states.

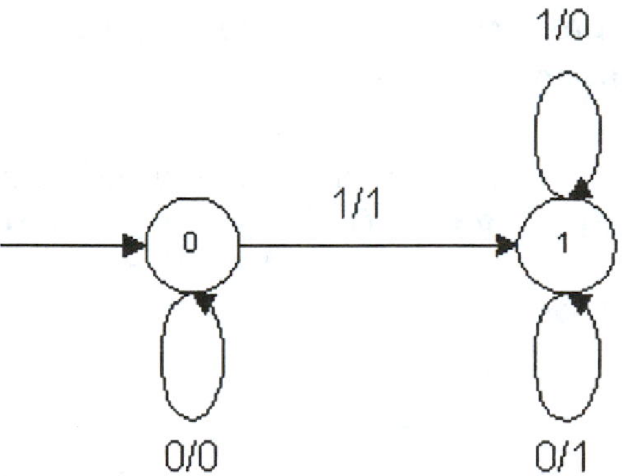

Figure 22.1: State diagram

The format of the labels on each transition are *in/out*, the machine's current input and output.

Procedure

1. Load the circuit **E22-1**, shown in Figure 22.2.

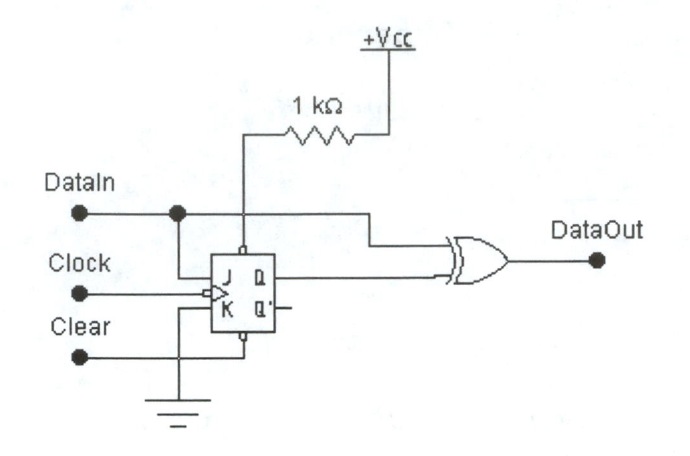

Figure 22.2: Serial 2's complementer

The state diagram in Figure 22.1 is implemented by the state machine in E22-1. This circuit takes a serial bit stream representing a multibit binary number (LSB first) and generates its 2's complement. For example, the binary number 10101100 has a 2's complement of 01010100. If the 10101100 pattern is input to E22-1 (LSB first, MSB last), the 01010100 pattern is output (LSB first as well).

The 2's complement is accomplished by passing all zero bits and the first one bit through without change, then passing the opposite of each bit for the remainder of the input pattern. Add the components required to verify this technique during simulation.

2. The state diagram for a state machine designed to recognize any binary number evenly divisible by four (modulo-4 detector) is shown in Figure 22.3.

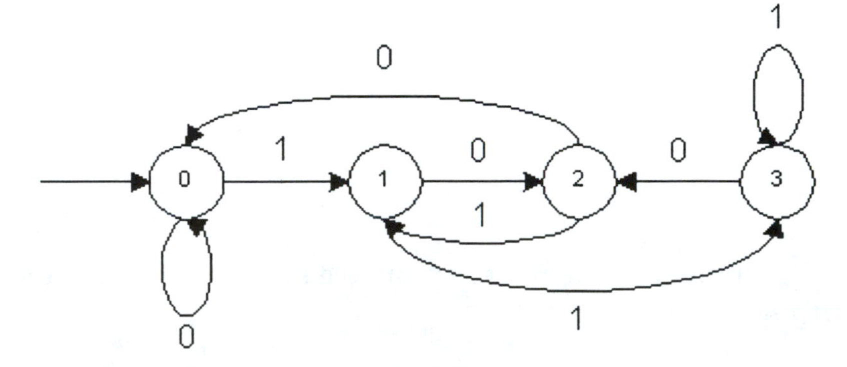

Figure 22.3: State diagram for modulo-4 detector

A sample number evenly divisible by four is 10011101000100 (10,052 decimal, which equals 2,513 times 4). Beginning with the MSB as the first input bit, trace the transitions in the state diagram. Begin at state 0. If you end up in state 0 after entering the last bit, the number is evenly divisible.

3. Load the circuit **E22-2**, shown in Figure 22.4.

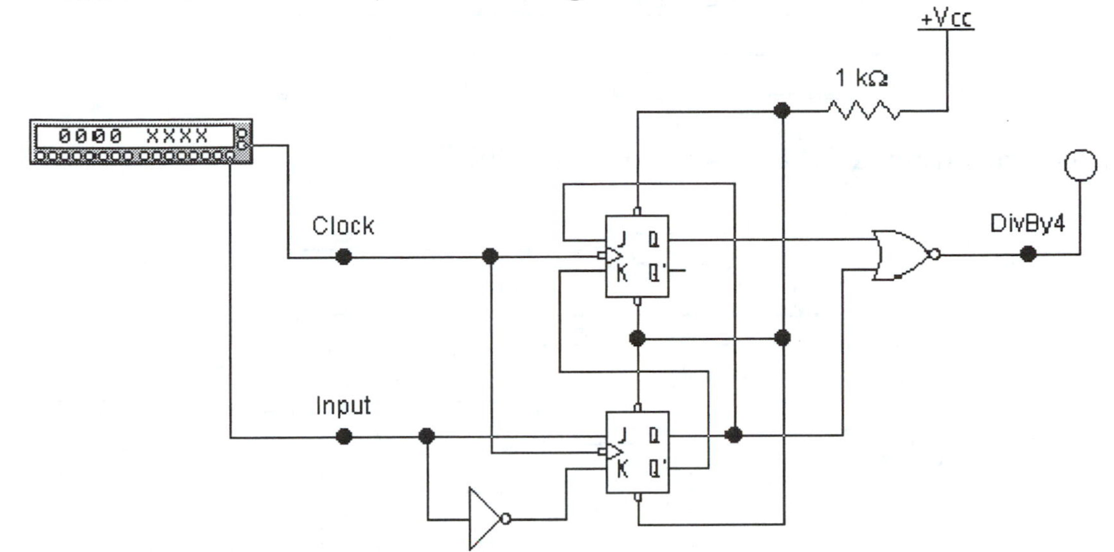

Figure 22.4: Modulo-4 detector

If the modulo-4 detector ends up in state 0, the NOR gate turns the logic indicator on.

4. Figure 22.5 indicates the test pattern saved in the Word Generator.

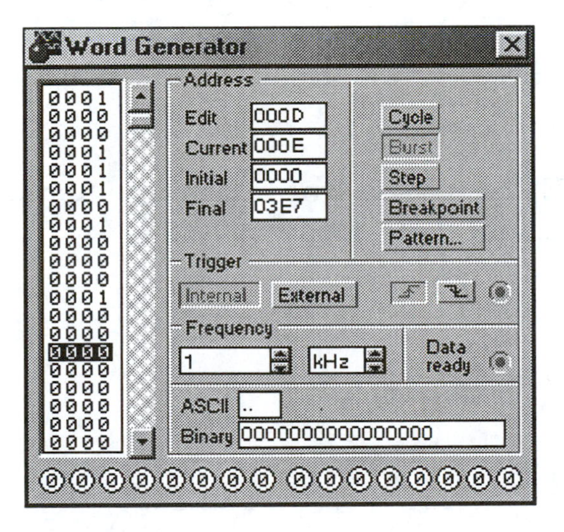

Figure 22.5: Test pattern 10011101000100

The line containing the last zero bit (line 000D) is set as a breakpoint. Simulate the circuit and verify that the number is recognized.

5. Choose two additional numbers, one divisible by four and the other not. Process them with the modulo-4 detector. What are the results?

INPUT NUMBERS	RECOGNIZED?

6. Load the circuit **E22-3**, shown in Figure 22.6.

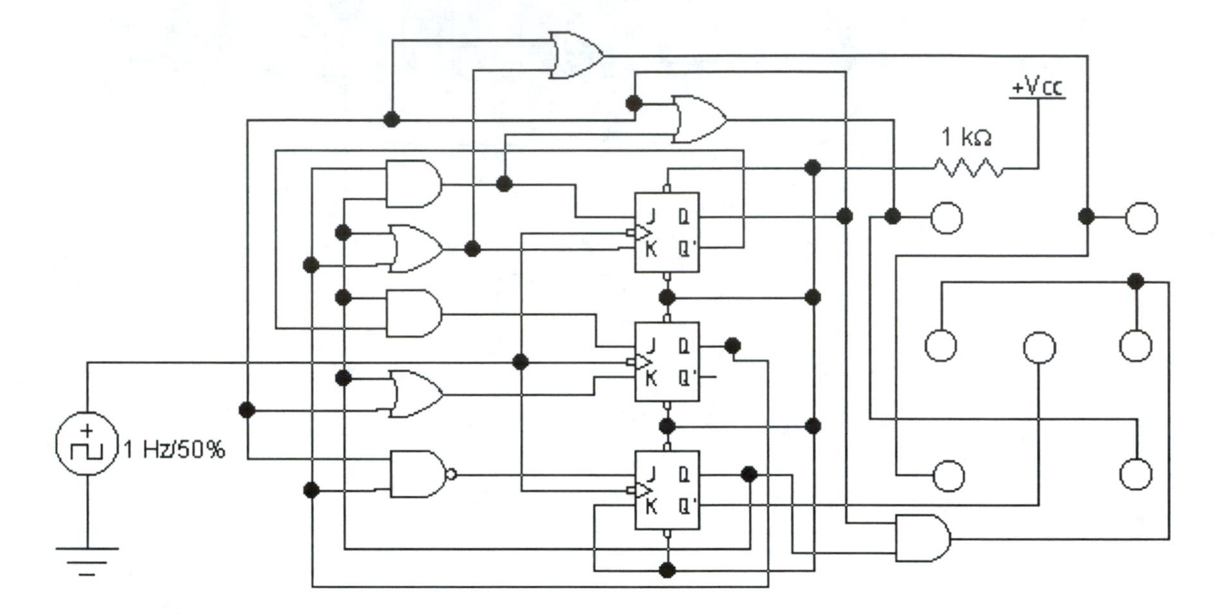

Figure 22.6: Digital dice

Simulate the circuit. How many states are there? What are the 3-bit state values?

7. Redesign circuit E22-3 using a synchronous counter. This will eliminate the flip flops and many of the support gates.

8. Examine the state diagram in Figure 22.7.

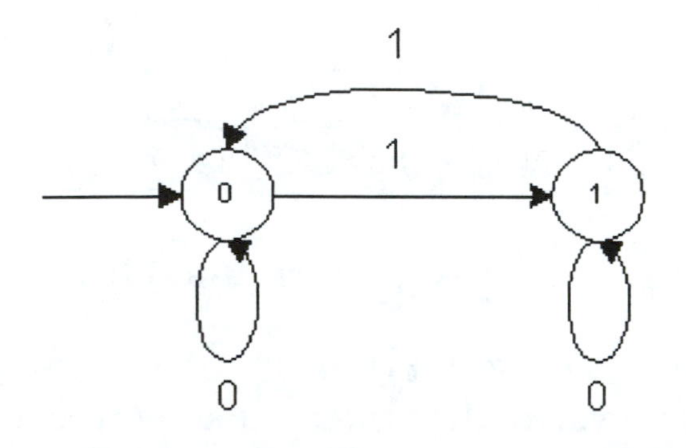

Figure 22.7: Parity detector

Show how a D-type flip flop could implement this state diagram.

9. Which state represents even parity?

10. *Troubleshooting:* Load the circuit **E22-4**, shown in Figure 22.8.

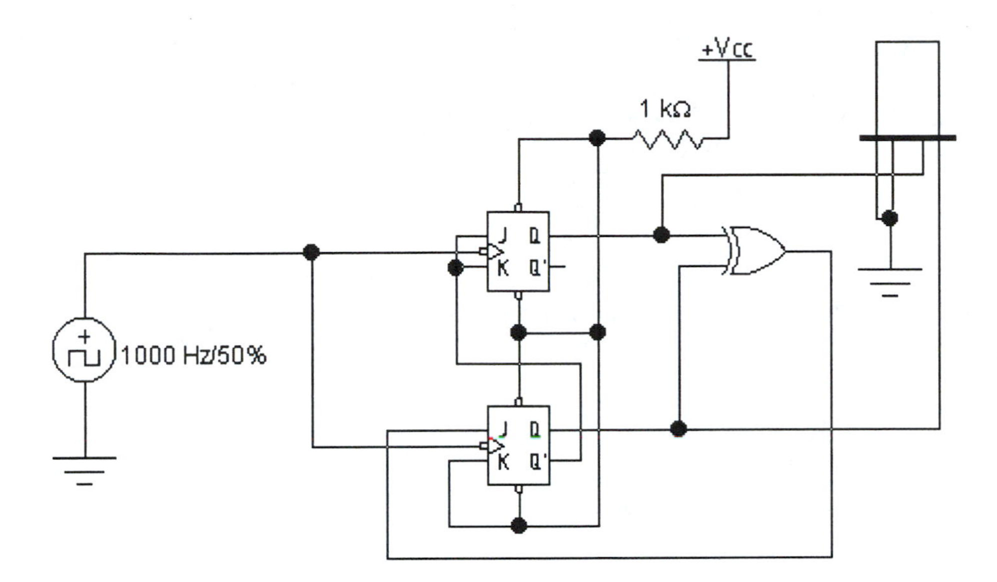

Figure 22.8: Troubleshooting circuit

The count sequence on the display should be 2, 1, 0, 2, 1, 0, etc. The circuit is not working properly. Determine the J and K logic levels assuming an initial 00 pattern on both Q outputs. Then repeat for each state pattern and use your results to find the problem.

STATE	J	K
00		
01		
10		
11		

Discussion

While reviewing your data and results, provide detailed answers to each of the following:

1. Explain how the serial 2's complementer circuit implements its state diagram. Assume the flip flop starts with Q equal to zero.

2. What are the states visited during processing of the 10011101000100 pattern in step 2?

3. How might a modulo-3 detector be constructed?

4. Which logic gates in the digital dice circuit are used to guide state transitions?

Experiment 23

Synchronous Counters

Name ______________________ **Date** __________

Objectives

The objectives of this experiment are to:

- Compare ripple counters with synchronous counters.
- Examine the operation of synchronous counters.

Introduction

One useful application of synchronous logic is in the world of counters. Synchronous counters can be clocked faster than ripple counters and have the added advantage that all outputs become valid at the same time, rather than after an accumulated ripple delay. In this experiment we will examine several different synchronous counters.

Procedure

1. Load the circuit **E23-1**, shown in Figure 23.1.

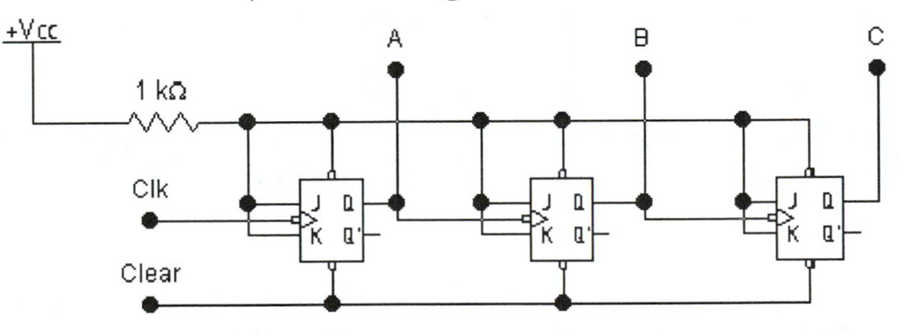

Figure 23.1: 3-bit binary ripple counter

Notice several characteristics about the ripple counter:

- Each flip flop is set up for toggle mode.
- The Q output of one flip flop is the clock for the next (therefore, all flip flops are not clocked at the same time).
- The counter responds to a negative clock edge.

2. Load the circuit **E23-2**, shown in Figure 23.2.

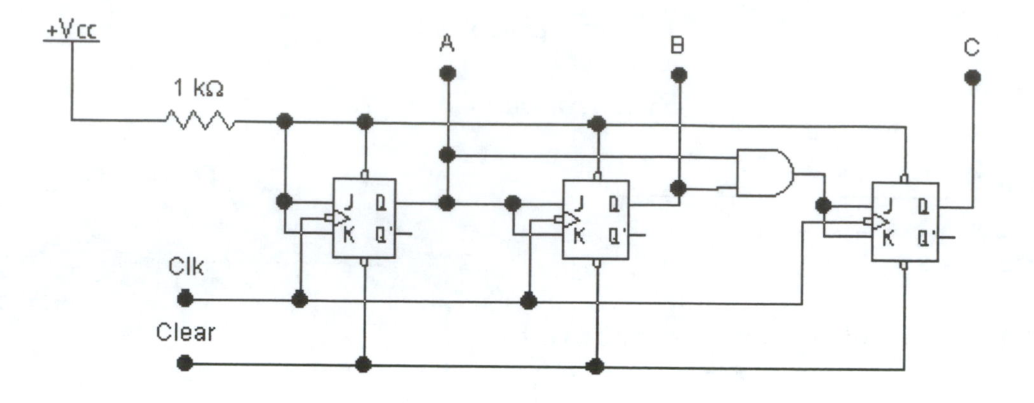

Figure 23.2: 3-bit synchronous counter

Note the differences between the 3-bit synchronous counter and the 3-bit ripple counter:

- All flip flops are clocked at the same time (which, in turn, guarantees that all outputs change at the same time, rather than after various ripple delays).
- Though the first flip flop is wired for toggle, the second and third flip flops may toggle, or may *stay in the same state*, depending on whether J and K are both high or both low.

3. Add the components required to test the synchronous counter and verify that it counts from 000 to 111, rolling over to 000 on the eighth clock pulse.

4. Using the tools of synchronous logic design (state diagram, state transition table, excitation table), develop and test a 4-bit synchronous counter.

5. Load the circuit **E23-3**, shown in Figure 23.3. The 74160 is a synchronous decade counter. Simulate the circuit. Does it count from zero to nine? What do you notice about the logic indicator connected to the RCO output?

6. Connect the Logic Analyzer so that the clock, the A through D outputs, and the RCO output can be examined. Exactly what is the relationship between all three?

7. Right-click on the 74160 and select Help to examine its truth table. What are the A, B, C, and D inputs for? When are they used?

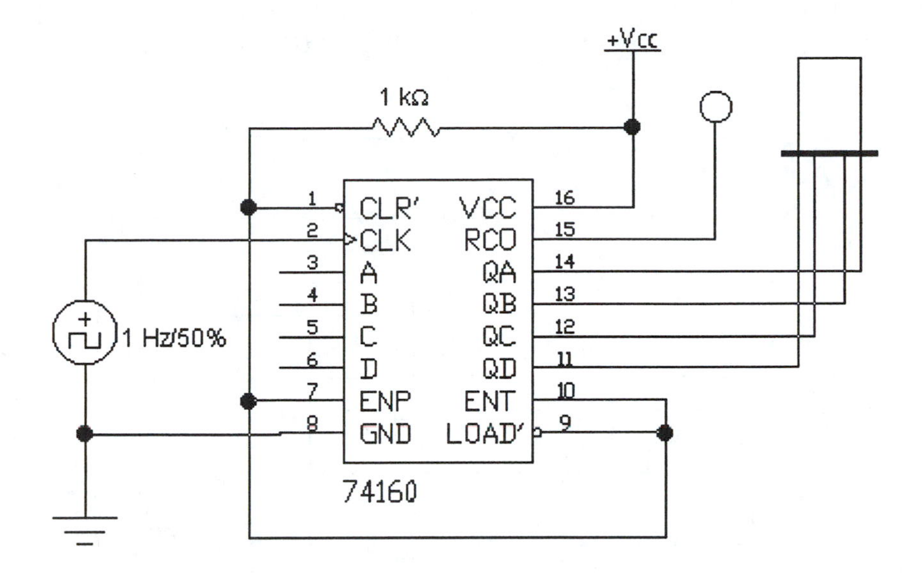

Figure 23.3: 74160 4-bit synchronous counter

8. Load the circuit **E23-4**, shown in Figure 23.4.

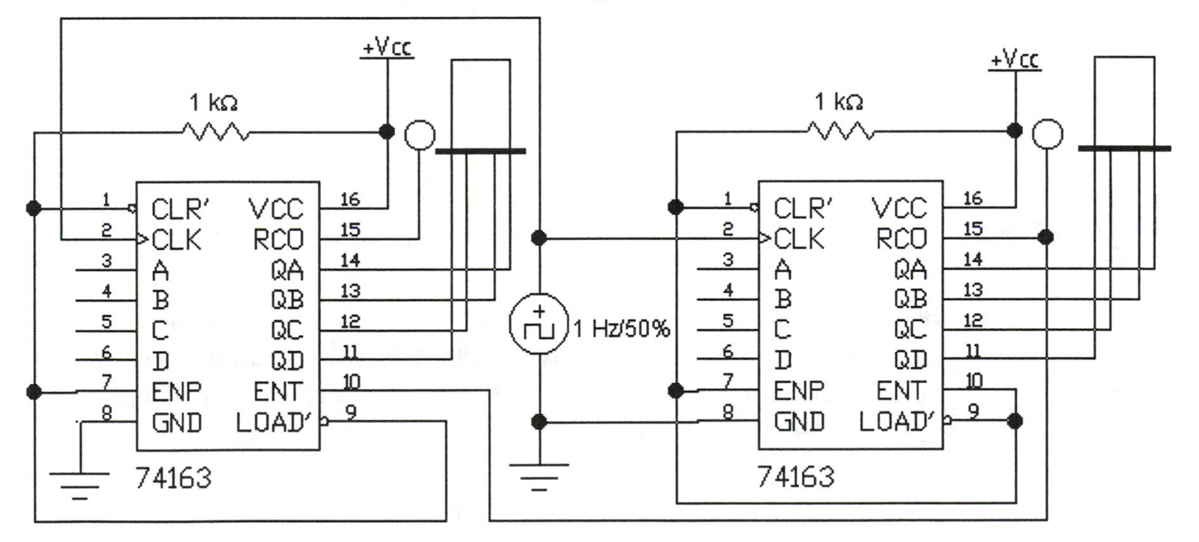

Figure 23.4: Cascaded synchronous counters

Notice that both counters are clocked at the same time. The RCO output of the lower four-bit counter feeds the ENT input of the upper four-bit counter. This connection provides the cascade feature.

9. Simulate the circuit. When does the upper four-bit counter increment?

10. Load the circuit **E23-5**, shown in Figure 23.5.

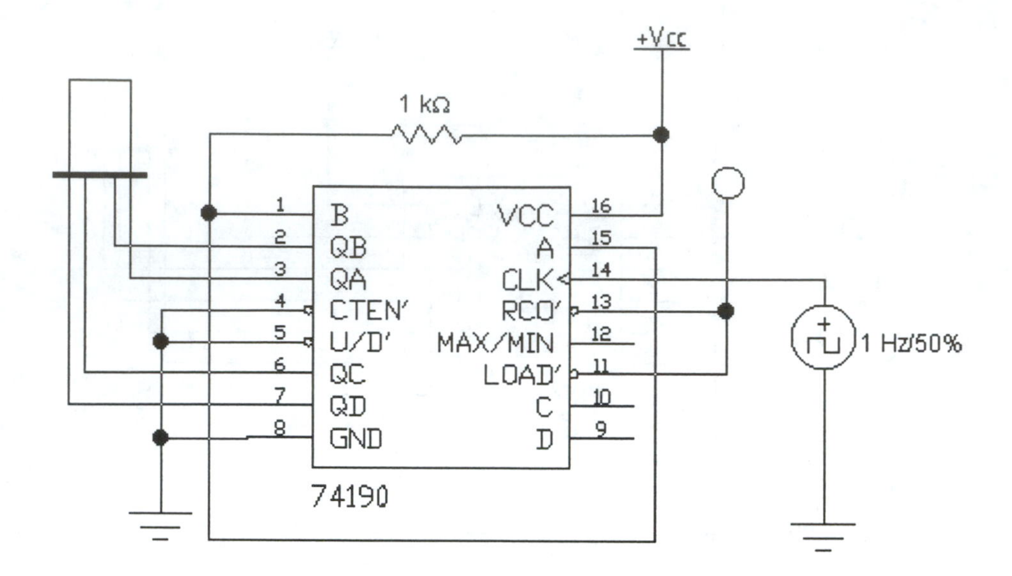

Figure 23.5: Synchronous modulo-6 counter

Verify that there are six states in the synchronous modulo-6 counter. Record the states indicated by the display.

0	1	2	3	4	5

11. Modify the circuit so that it counts from zero to five and resets on six.

12. Use two 74190's to build a modulo-100 counter. Do not use any logic to detect a specific output pattern. The counters should be loaded with an initial pattern that causes them both to roll over to zero on the 100th clock pulse. Wire the counters for up counting.

13. Repeat step 12 with the counters set for down counting.

14. Design a circuit that begins its count at 0011, counts up to 0111, then changes direction and counts back down to 0011.

15. Examine Figure 23.6. This circuit is called a *ring* counter. Its output sequence is 1000, 0100, 0010, 0001 due to the shift register action. Explain why the ring counter is a synchronous counter.

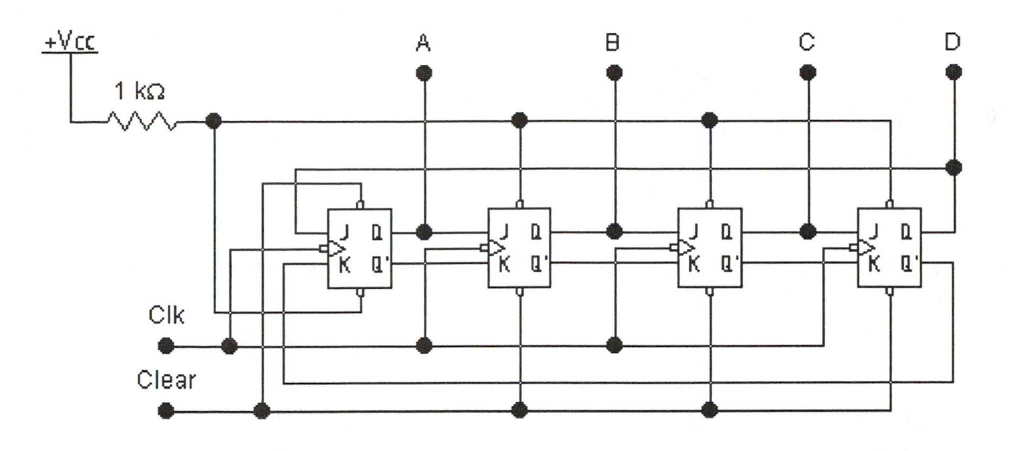

Figure 23.6: Ring counter

16. *Troubleshooting:* Load the circuit **E23-6**, shown in Figure 23.7.

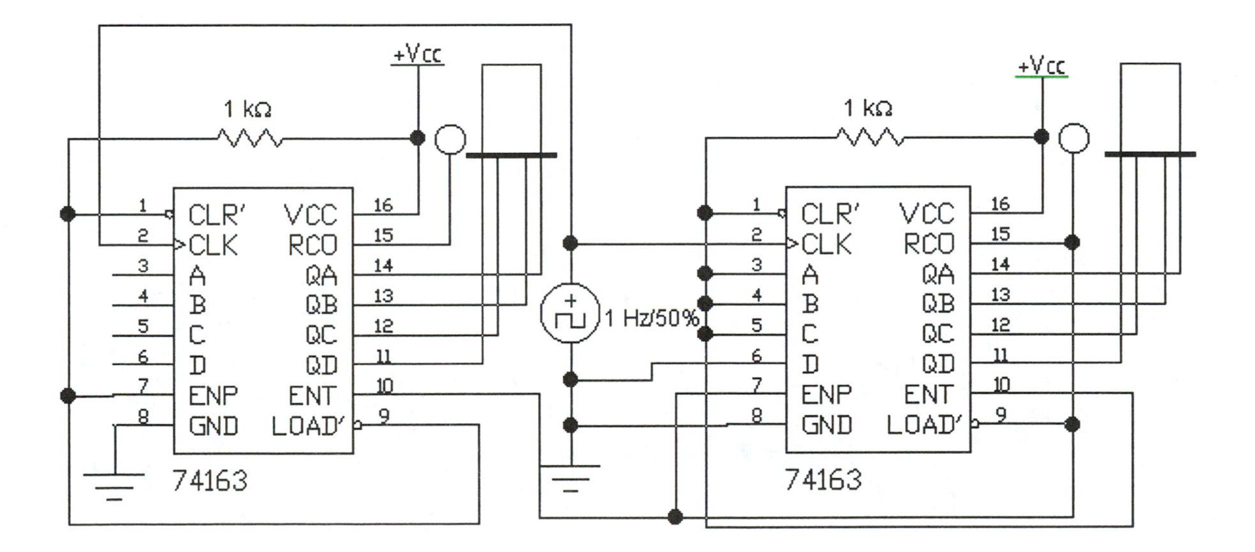

Figure 23.7: Troubleshooting circuit

This circuit is similar to E23-4, except it has been modified to begin the lower 4-bit counter at 0111 instead of 0000. It may not have been modified correctly. Verify its operation and make any necessary changes.

Discussion

While reviewing your data and results, provide detailed answers to each of the following:

1. Explain why a synchronous counter can be clocked faster than a ripple counter.

2. What is the purpose of the RCO output?

3. Explain how a synchronous down counter can be used as a modulo-6 counter.

4. Explain how a synchronous counter could be used as the counter for the hours digit in a digital clock.

Experiment 24

The Digital Vending Machine

Name ______________________ **Date** __________

Objectives

The objective of this experiment is to:

- Examine the operation of a simple digital vending machine.

Introduction

The simple digital vending machine examined in this experiment accepts nickels and dimes (up to nine each) and turns on $0.20 and $0.45 indicators when enough money has been entered.

Procedure

1. Load the circuit **E24-1**, shown in Figure 24.1.

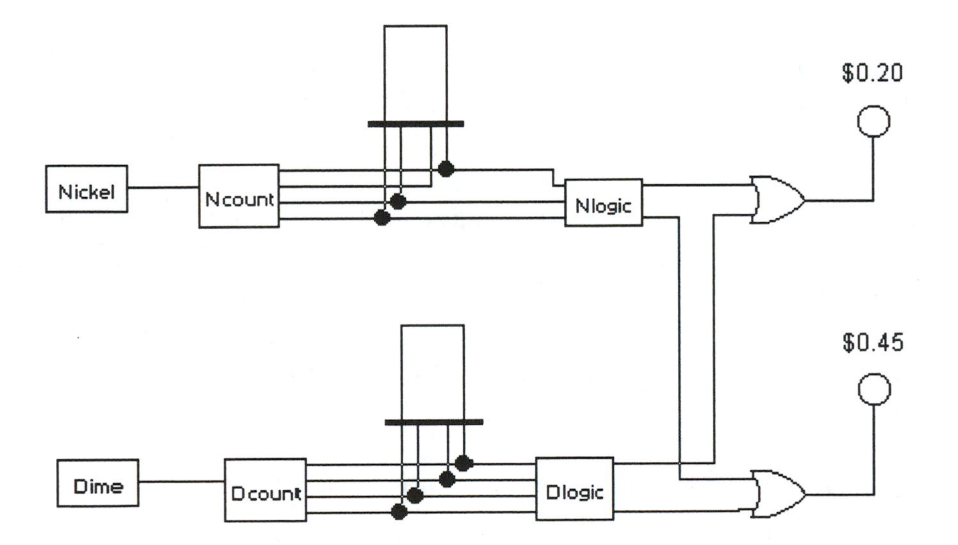

Figure 24.1: Vending machine logic

The Nickel and Dime subcircuits contain switches ('N' and 'D') that represent the coin mechanism. Either switch must be opened and then closed (two keypresses in a row) to register a coin. The displays indicate the number of coins entered and roll over from nine to zero, effectively stealing the money. The Nlogic and Dlogic subcircuits contain combinational logic that decodes when enough coins have been entered to buy a $0.20 or $0.45 item.

2. Simulate the circuit. Both displays should read zero. Enter nickels one at a time. Do the $0.20 and $0.45 indicators come on at the correct time?

3. Resimulate, this time entering only dimes.

4. Resimulate and enter one dime and two nickels. Is the circuit working properly? The answer should be no. A third subcircuit is required that combines the number of nickels *and* dimes already entered. Design the subcircuit and add it to the original machine. *Hint:* two nickels make one dime.

5. Test your design with the following coin combinations:

- One dime, two nickels
- Two dimes, five nickels
- Three dimes, three nickels
- Four dimes, one nickel

6. Change the 7490 counters in the Ncount and Dcount subcircuits to 4-bit binary synchronous up/down counters.

7. Create two additional subcircuits similar to the Nickel and Dime subcircuits. These will be the Buy20 and Buy45 subcircuits. After the user has entered enough money, these subcircuits should allow a product to be purchased.

8. Once a product is purchased, the nickel and dime counters must be decremented, so that only the change is left on the display. For example, if five dimes are entered and Buy20 is selected, the final dime display should read three. Design a subcircuit that will decrement either counter from 1 to 9 times. A 4-bit input specifies the number of times to decrement the counter.

9. What circuitry is required to add a $1.00 product?

10. *Troubleshooting:* Examine the state diagram in Figure 24.2.

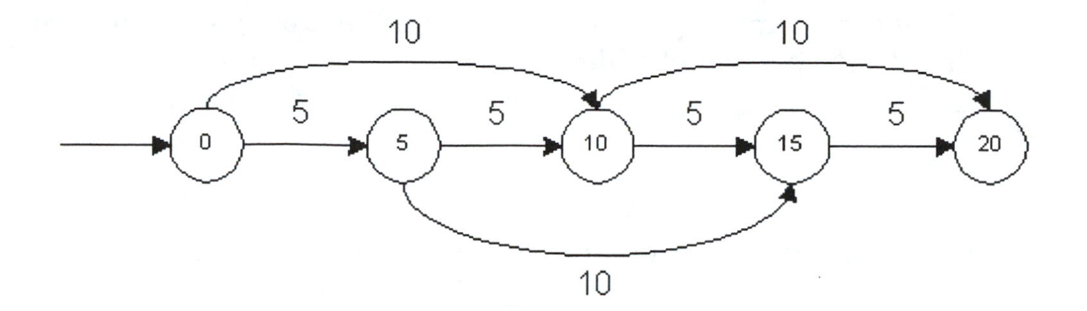

Figure 24.2: $0.20 state diagram

Ignoring the problem associated with entering a third dime, there are several ways to get to the 20 state. Does the circuit **E24-2**, shown in Figure 24.3, implement this state diagram?

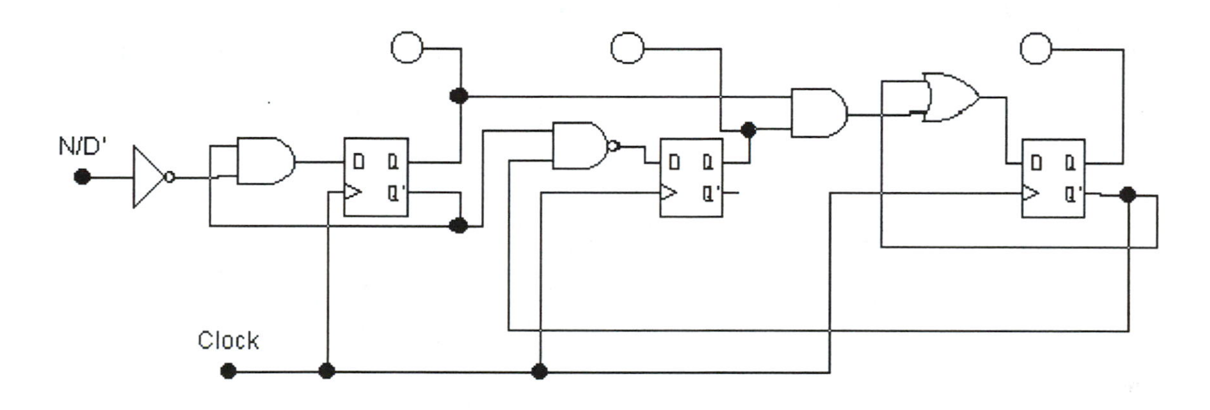

Figure 24.3: $0.20 state machine

Discussion

While reviewing your data and results, provide detailed answers to each of the following:

1. Explain how the Nlogic subcircuit works.

2. Explain how the Dlogic subcircuit works.

3. Explain how the complexity of the money decoding and change logic could be simplified with a parallel adder.

4. How many states would be required by a state machine allowing a maximum of eight nickels and five dimes?

Experiment 25

Generating a Digital Waveform

Name ______________________ **Date** __________

Objectives

The objective of this experiment is to:

- Examine a technique for generating digital waveforms.

Introduction

Figure 25.1 shows the timing of a simple digital waveform. Each major horizontal division represents one bit. The entire waveform requires 16 bits, in the following order: 1010000010111111.

Figure 25.1: Sample digital waveform

The waveform may represent the control signal to a value on an assembly line, the pattern transmitted by the 'Play' button on a remote control, or the serial transmission code for an ASCII character.

In this experiment we will examine one method of designing a digital circuit capable of generating the waveform in Figure 25.1 (or any other digital waveform).

Procedure

1. Load the circuit **E25-1**, shown in Figure 25.2.

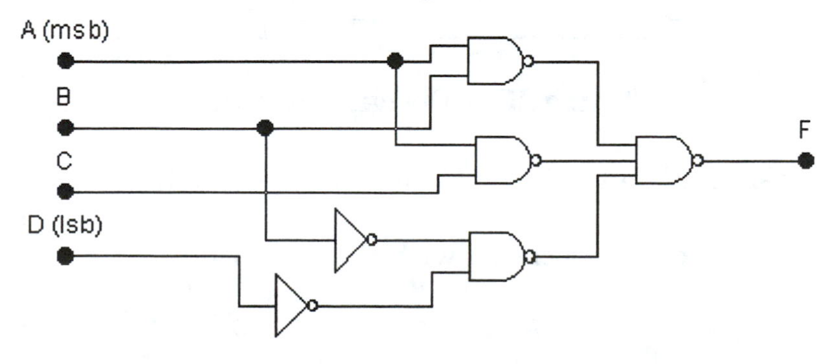

Figure 25.2: Waveform generator circuit

2. Use the Word Generator to drive the four inputs. Load the Up counter pattern into the Word Generator and let it cycle. This will apply a 4-bit count to the inputs, from 0000 to 1111, stepping through each bit of the waveform in sequence. View the waveform with the Logic Analyzer and verify that it looks like Figure 25.1 (over and over).

3. Remove the Word Generator and load a 7493 4-bit binary counter from the 74xx Template parts bin. Connect the counter so that it drives the four inputs the same way as the Word Generator. Note: Connect the LSB of the counter (Q_A) to the LSB of the waveform generator circuit (D), and so on. Verify that the same waveform is generated.

4. Enter the 16-bit waveform data 1010000010111111 into the Karnaugh map shown in Table 25.1. Solve the map and convert the result into all-NAND logic using DeMorgan's Theorem. Compare your resulting expression with the logic of the waveform generator circuit.

	CD	CD	CD	CD
AB	00	01	11	10
00				
01				
11				
10				

Table 25.1: Karnaugh map for waveform generator circuit

5. Load the 16-bit waveform pattern into the Logic Converter and solve for the simplified equation. Is it the same as the one you found in the Karnaugh map?

6. Use a Karnaugh map to design the waveform shown in Figure 25.3.

Figure 25.3: Design waveform

7. Test your design using the Word Generator and the Logic Analyzer.

8. What is the waveform associated with this equation:

```
F = A'BC + BD + AC'D
```

Use the Logic Converter, a Karnaugh map, or an actual circuit to determine the binary pattern. Write each bit in its associated slot in Table 25.2.

0	1	2	3	4	5	6	7	8	9	10	11	12	13	14	15

Table 25.2: Bit pattern for waveform equation

9. *Troubleshooting:* Load the circuit **E25-2**, shown in Figure 25.4. The desired waveform for this circuit is shown in Figure 25.5, but is not being correctly generated. Determine what is wrong with the circuit.

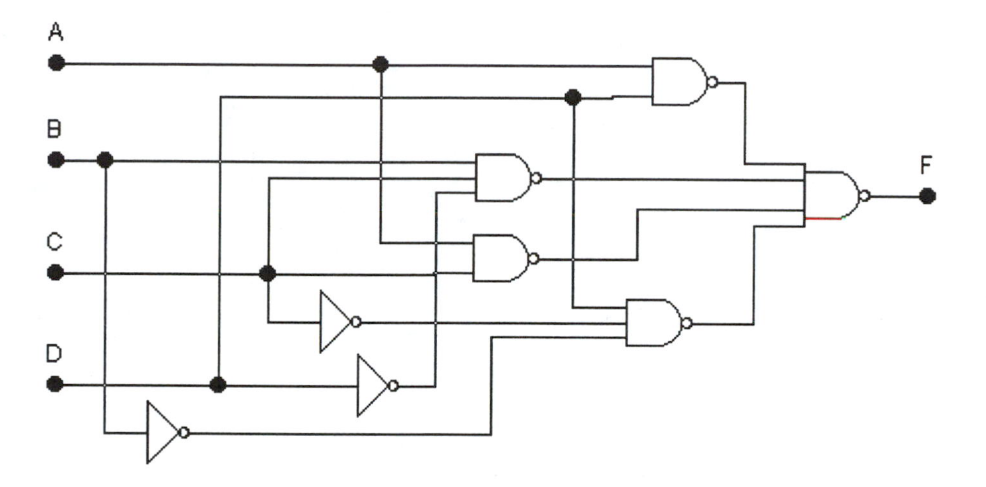

Figure 25.4: Troubleshooting circuit

Figure 25.5: Desired troubleshooting waveform

Discussion

While reviewing your data and results, provide detailed answers to each of the following:

1. Is it possible to represent the waveforms in Figures 25.1 and 25.3 with fewer bits?

2. When using the 7493 counter, why not connect Q_A to A (of the waveform generator), Q_B to B, and so on?

3. How many inputs are required for a waveform generator, if the waveform contains 37 bit times? What about 80 bit times?

4. What is the waveform produced by this equation: $F = AB + BC + CD$

Appendix: Electronics Workbench Reference

This appendix is intended as a quick reference for typical operations performed in Electronics Workbench (EWB), as well as a pictorial guide to where things are located.

The Main Window

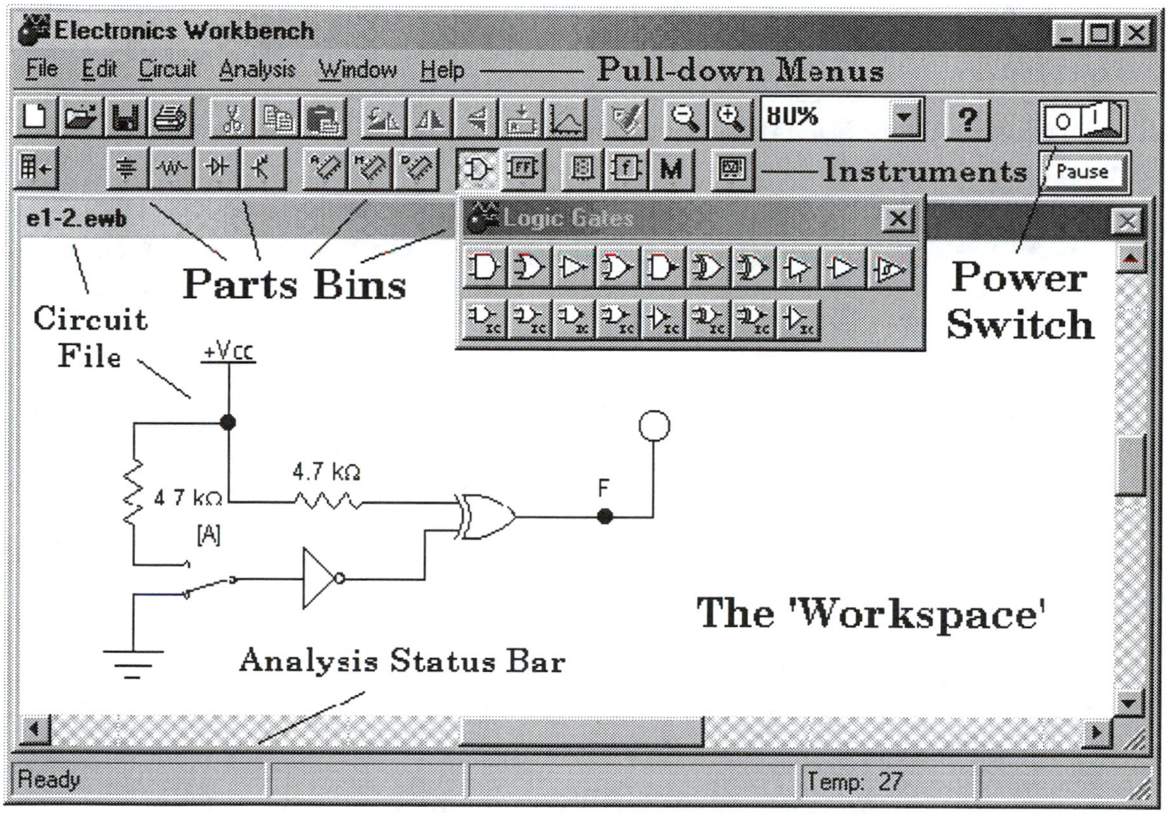

Figure A.1 shows a sample screen shot of Electronics Workbench in operation. Labels have been added to identify the major components.

Figure A.1: EWB main window

The workspace is where all your work is performed. Circuits are constructed by dragging components from the selected parts bin into the workspace. Instruments are selected by dragging them 'off the shelf' and into the workspace. Components are wired together by dragging wires from one component lead to another. Entire circuits can be loaded or saved using the File menu. Both DC and AC analysis is performed. Analog and digital circuits may be mixed together. The pull-down menus, types of instruments, and the contents of each parts bin are detailed in the next three sections.

The Pull-down Menus

There are six pull-down menus: File, Edit, Circuit, Analysis, Window, and Help. Left-click to select a menu, or press ALT followed by the underlined letter in the menu name. Left-click on the desired option to choose it. Left-click anywhere in the circuit window to close the menu without choosing an option.

The Instruments

Electronics Workbench provides seven virtual instruments for making measurements and generating signals. Figure A.2 identifies each instrument. To use an instrument, drag it from the 'shelf' into the workspace. Then connect to the instrument terminals as you do any other connection. To see the instrument's details (controls, displays, connector names), double-click on the instrument to open it. To remove an instrument from the workspace, simply delete it.

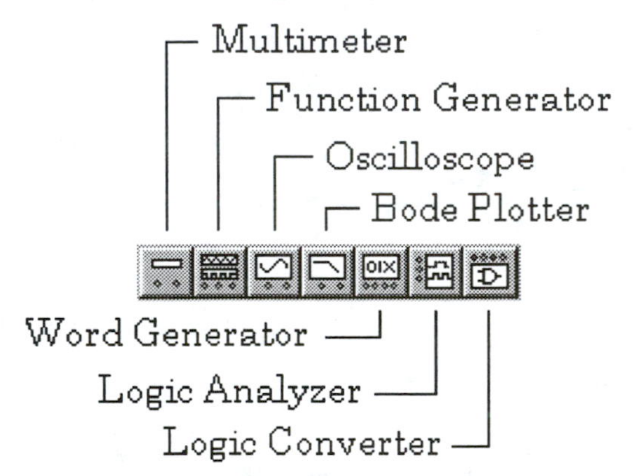

Figure A.2: EWB instruments

The Parts Bins

Electronics Workbench provides twelve different parts bins for building your circuits (as shown in Figure A.3). In order, they are: Sources, Basic, Diodes, Transistors, Analog IC's, Mixed IC's, Digital IC's, Logic Gates, Digital, Indicators, Controls, and Miscellaneous. To select a bin, click on it once. To move a part into the workspace, drag it from the parts bin to the workspace. To delete a part, click on it once and press 'Delete'.

Figure A.3: Parts bins